Confronting Climate Change with Indigenous Wisdom and Western Science

PROMPT

Glen S. Aikenhead

Confronting Climate Change with Indigenous Wisdom and Western Science

The Sustainability Revolution

PETER LANG

New York - Berlin - Bruxelles - Chennai - Lausanne - Oxford

Library of Congress Cataloging-in-Publication Control Number: 2024031387

Bibliographic information published by the Deutsche Nationalbibliothek.
The German National Library lists this publication in the German National Bibliography; detailed bibliographic data is available on the Internet at http://dnb.d-nb.de.

Cover design by Peter Lang Group AG

ISBN 9781636678054 (hardback)
ISBN 9781636678061 (ebook)
ISBN 9781636678078 (epub)
DOI 10.3726/b21525

Published by Peter Lang Publishing Inc., New York, USA

info@peterlang.com – www.peterlang.com

This publication has been peer reviewed.

Dedication

This book is dedicated to three colleagues/friends whose constant support and mentoring over many years have enriched my academic and personal lives immeasurably: Dr. Leo Klopfer (deceased, Dec. 24, 2022); Dr. Marie Battiste (Mi´kmaw); and Elder Danny Musqua (Nakawē, deceased Dec. 5, 2022).

In a circle, nothing really dies out; things might get old and wear away but they renew again, generation after generation. That is what the circle is about.—Danny Musqua

The purpose for our existence is to learn; to acquire knowledge around us through our hearts, minds, bodies, and spirits, in a balanced way.
—Danny Musqua

CONTENTS

FOREWORD

For those who believe, no explanation is necessary; for those who don't believe, no explanation is possible. But in a non-binary world, inhabited by humans like us, there should be something for those who feel they should believe in climate change but who need more explanation. And here, in Aikenhead's book, Glen provides that explanation with evidence substantiated by science. Not just through Western science. Much of the evidence is coming from diverse Indigenous knowledges from all over the world— through stories passed down from generation to generation, through dance, through art, through song—by knowledgeable people who want to pass on to their children what is important for their lives. Over millennia, now, and into the future.

And they are willing to tell us, to share with us as well, if we are willing to listen and make proper use of their knowledge.

Glen Aikenhead has listened and learned from Indigenous people in a variety of places, and in this book, he has merged Western science and Indigenous ways of living in nature in a way we can all appreciate. We are sure this book will help us as we all are *Confronting Climate Change*.

Dr Michael Michie
Darwin, Australia
March, 2024.

Dr Michelle Hogue
Lethbridge, Canada

PREFACE

Modern science came about due to a revolution (i.e., "the scientific revolution," 1543–1660 CE) and eventually became what science is today—one of the weapons to fight against the current climate-change crisis. The revolutionary transition took place during a period of time known as the European Renaissance (1300–1700 CE about). Some European cultural beliefs were changing from ancient Greek thinking to modern Renaissance thinking. The ancient Greek thoughts were accepted as true if they came out of the pure minds of Plato or Aristotle. That was enough to make them true.

Those two Greek philosophers stated that planets moved across the night sky for the same reason the Sun did during the day. Because they are perfect heavenly bodies, they must travel in perfect circles around the Earth, and always with the same speed.

Some natural philosophers spread Copernicus' radical thoughts about the Earth moving around the sun. It caused some people to become so angry that they burned some of those radical thinkers at the stake.

Johannes Kepler was a natural philosopher (early scientist) and a superb mathematician who was setting aside the ancient Greek way of deciding what or who should be believed, concerning the physical universe. The Greek way was to rely on whatever the purity of Plato's or Aristotle's minds dictated. The revolutionaries, such as Kepler, Galileo and Newton, wanted to reason with evidence on their own.

European Renaissance culture of natural philosophers, like Johannes Kepler, adopted the view that *valid evidence* supporting an idea would determine what or who to believe. European culture altered accordingly. This issue of who to believe continues today concerning pandemics and climate change. The book, *Confronting Climate Change*, confirms the above by tracking the evidence for a change in culture that evolves from the same questions: How do I decide? and Who do we believe, about climate change?

The book's subtitle *answers* this compound question: a sustainability revolution. The historical summary just above looks like a gross overgeneralization without the dramatic details waiting for you to read in the book's chapters along with its Epilogue. Its conclusion will be illuminated by the collaboration among:

- people who by tradition or by habit generally live a sustainability culture; and
- broadly defined: science, economics, engineering, environmental and political-science communities; all motivated to get rid of the climate-change crisis.

All of these people need an international leader or group of leaders with a plan and expertise to rid the Earth of the threat. Keep your eyes on Dr. Mark Carney, Dr. David Suzuki, and other potential nominees as mentioned in the book.

Readers can expect to acquire clear ideas about how scientists do their *normal* work and how they interact with fellow scientists. This type of content, of cource, is known as "*the nature of science*" (NOS). Another type is how the scientific enterprise interacts with society, an activity that scientists now call "post-normal science"[1] or "socio-scientific issues" (SSI).[2] The book points out for the reader when this knowledge is used, mostly at the climate-change crisis.

Accordingly, the book has two goals for readers, with implications for their everyday lives. First, to understand the nature of science, scientists and society well enough to challenge those people who: (a) misrepresent science or scientists; (b) dismiss legitimate scientists' professional research results; (c) demonstrate the "Dunning-Kruger Effect;"[3] (d) cannot distinguish between the latest

1 Funtowicz & Ravetz (1993)
2 Klopfer & Aikenhead (2022)
3 Duignan (n.d.)

scientific knowledge and an outsider's interpretation of it; or (e) engage in other tactics such as threatening violence.[4] Such people are characterized by the established news media as the anti-science crowd, or the I-am-the-centre-of-the-universe crowd.

The overall argumentation of the book draws on Kepler's rich concept of harmony to conclude that today's Indigenous wisdom has a role to play in helping end the climate-change crisis by collaborating with scientists, economists, engineers, political leaders and the general public.

The accompanying culture change will bring about a sustainability revolution. But sustainability is not a strength in today's world. The world requires a much stronger pulse of sustainability.

4 Lomborg (2022)

ACKNOWLEDGMENTS

These acknowledgments require two sections to accurately represent the people to whom I am deeply indebted, outside of my incredibly patient and understanding wife, Brenda: (a) Dr. Leo Klopfer; and (b) a host of Indigenous colleagues and friends, the majority being keepers of Turtle Island (North America).

Part 1. Dr. Leo Klopfer

Leo was a highly respected and productive American university science education professor, who is best known for his editorship (1979–1992) of the acclaimed research and policy journal *Science Education*.

Coincidently, he and I had experienced parallel university education backgrounds:

- *BSc Chemistry*: Klopfer (1957, Univ. of Chicago); Aikenhead (1965, Univ. of Calgary)
- *Master of Arts in Teaching*: both at Harvard Univ.: Klopfer (1959,); Aikenhead (1966)
- *Doctorate of Science Education*: Harvard Univ.: Klopfer (1962); Aikenhead (1972).

You need to know that in the 1940s, J. B. Conant, President of Harvard, challenged science students to understand their subject *more thoroughly and deeper* than it is normally taught as conventional abstract subject matter. When they become adults, he thought science students should be able to interact with the general public on science-related matters by drawing on its features that historians, philosophers and sociologists have discovered about science.

This knowledge genre is called: "*the nature of science*" (NOS)[56] and includes *socio-scientific-issues* (SSI).[7] Leo also developed a student instrument, entitled "Test on Understanding Science" (TOUS). It assessed students' understanding of NOS and SSI content.

In Leo Klopfer's doctorial program, he pioneered the development and publication of six *history of science* case studies for high school students; for example: *The Cells of Life*, and *The Discovery of the Halogen Elements*.[8] Students authentically repeated some of the experiments carried out originally. It gave them the feeling of taking part in the advancement of science.[9]

Leo and I finally met face-to-face at a research conference, two years after I became employed as a science education professor at the University of Saskatchewan in 1971. We had much to talk about. He apparently decided to be a mentor to my science education career, for which I became forever grateful and indebted. This is the reason for this first section appearing in the book's Acknowledgment. And of course, the two of us became strong friends.

After he retired in 1994 from Pittsburgh University, our connections gradually waned. However, we rekindled our friendship in 2020 when he was 91 years old. It began by enjoying monthly phone calls. He felt uneasy using email.

Suddenly he became hospitalized at a retirement facility in the Philadelphia area. He was kept alive by an exhausting medical procedure three days a week. As a result, Sunday morning phone calls became our time for conversations, if he was not too tired.

When reminiscing about our past work he talked me into co-authoring an article on the history of humanistic science education, an expression that goes back to 1971. Leo was not well enough to do any research or writing, but he had not lost his talent for mentoring my writing efforts. Our article was published[10]

5 Aikenhead (2007)
6 Lederman (2007)
7 Klopfer & Aikenhead 2022
8 Klopfer & Cooley (1963)
9 Aikenhead (1979)
10 Klopfer & Aikenhead (2022)

in volume 3 of the 2022 *Science Education*. The whole process brought him back to reliving his Editorship of the journal *Science Education*. It brightened some of his days.

I was thoroughly rewarded by being able to repay him, in part, for the wonderful mentoring he had done for me in the 1970s and 1980s when my career began taking shape.

During one of our January 2021 phone calls, he exclaimed, "We should write a trilogy about the Scientific Revolution, one book each for Kepler, Galileo, and Newton." I talked him into one book at a time. "Let's begin with Kepler, and see how that goes," I suggested. And as they say, the rest is history.

Sadly, Leo passed away on December 24, 2022, at the age of 93, after a long hospitalization for medical treatment that happily prolonged his life. My mission became to complete the book that we had started.

Happily, he vividly lives within me as he always will.

Part 2. Indigenous Colleagues and Friends

Tān´si. Glen nit´siyikāsan.
(Plains Cree for: "Hello, how are you? I am known as Glen")

With heart-felt pride, I acknowledge my Treaty 6 Canadian identity, living on nêhiyawaskîwin (Cree ancestral land). I wish to acknowledge all the kindness, generosity and patience shown me by Indigenous Elders, Knowledge Holders and students. Because of them, I have become a knowledgeable ally during the last 30 years of the following personal involvement.

My first extended conversation outside of the University of Saskatchewan with a First Nations or Métis person occurred in 1993 when I drove to Prince Albert, Saskatchewan, to take an Indigenous ethicist author to lunch. An essay by Willy Ermine had impressed me.[11] I wanted to ask him what research I could do that would be helpful to the Indigenous and Métis students enrolled in my science methods classes for the College of Education's Indigenous and Métis Teacher Education Programs. Willy patiently and kindly forgave my naïve question and simply advised me to get involved with the Indigenous community.

I was welcomed by the Saskatoon Tribal Council (STC), a cooperative of seven First Nations band associations in the vicinity of Saskatoon. They were about to initiate a "Super Saturday" program that bussed different school-aged

11 Eventually published in Ermine (1995)

groups of students from their reserve schools to the University Campus on designated Saturdays. Students became involved with some science-related activities with an Indigenous connection.

These activities were created by science, engineering, or medical professors (or by their designated grad students). Each Saturday's visit consisted of a talk from their school's Elder and the opportunity to share the land where the University sat. Both the activities and Elders created an emotional-based connection for students. I served on the STC's curriculum committee as their liaison with the university. I acknowledge the trust and support the STC afforded me.

As liaison, I attended culture sessions held by the University's Indigenous Elders who held circle seminars with the university faculty and grad students participating in Super Saturday. In these sessions, I learned from the Elders in much greater depth than from any excellent book I had read. I acknowledge their sharing with me Indigenous experiences and insights into Indigenous worldviews.

I became closely acquainted with Elder Danny Musqua, a member of the Bear Clan of the Saulteaux First Nation. We enjoyed many discussions, from which I gained wonderful insights. He became the Elder from whom I would always seek advice. He guided me in many ways during the years that followed the Super Saturday project. The book, *Bridging Cultures: Indigenous and Scientific Ways of Knowing Nature*[12] was solely dedicated to Elder Musqua. Sadly, on December 5, 2022, he found his star in the firmament.

I wish to acknowledge here the kindness, generosity, and patience of many Indigenous Elders, Knowledge Holders, and students. As a result of these experiences, I have become an informed strong ally over the past 30 years involved with:

1. *Local organizations*; for example: Saskatoon Tribal Council, Northern Lights School Division, Saskatoon Public Schools, the Federation of Sovereign Indigenous Nations, and the North East School Division.
2. *Elders*; for example: Danny Musqua (Nakawē), Leroy Little Bear (Kainai, Blackfoot Confederation), Henry Sanderson (Nehiyaw), Albert Scott (Nakawē), Darlene Speidel (Lakota), Oscar Kawagley (Yupiaq, Alaska), and Temi Nawi (Seediq, Taiwan).
3. *Knowledge Holders*; for example: Dr. Marie Battiste (Mi´kmaw, Canada); Dr. June George (West Indies Native); Waokiye Don Spiedel (Lakota,

12 Aikenhead & Michell, (2011)

Saskatchewan); Dr. Herman Michell (Nîhithaw, Sakstchewan); Dr. Elizabeth McKinley (Māori, Aotearoa New Zealand); Dr. Nancy Maryboy (Dené—Navajo, USA); Dr. Greg Cajete (Santa Clara Pueblo, USA); Brenda Green (Dakota, Saskatchewan); and Sharon Meyer (Nêhiyaw, Saskatchewan).

4. *Encounters*, such as: major conferences, workshops, part-time employment, and research projects. They facilitated my coming-to-understand Indigenous ways of understanding how to live in a good way.

My most significant experiences are condensed in what follows from here.

In 1993, Dr. Marie Battiste joined the U. of S. Education faculty and became a supportive colleague-friend for the next 15 years. For instance, she arranged for me to attend a long-weekend Turtle Island (North American) seminar about Indigenous understandings, held at the Banff (Alberta) School of Fine Arts. Very fortunately, I established a life-long colleagueship with the seminar leader, the eminent Elder Leroy Little Bear of the Kainai Nation in Alberta, and with his wife, Amethyst First-Rider. Future meetings became more engaging and productive.

Marie Battiste received funding for a 10-day 1996 international summer institute, "Cultural Restoration of Oppressed Indigenous Peoples." It attracted many delegates, including Leroy Little Bear and Amethyst First Rider, and several others whose enduring friendship has made them close colleagues of mine; especially: Dr. Greg Cajete, a Pueblo American Indian from New Mexico (*Native Science*);[13] Dr. Nancy Maryboy of the Diné (Navajo) Nation;[14] and Phil Duran, professor at the Northwest Indian College, Bellingham, Washington,[15] author of *The Condor and the Eagle*; that united the heart and mind in a way that created a new science worldview. All five of us have continued to share other conferences as well. They have been generous in sharing their Indigenous ways with me.

Similar encouragement came from the R&D carried out by the Alaska Native Knowledge Network for Yupiaq science students. Initiated by doctorial Yupiaq students exploring students' reactions to their students' Indigenous ways of living in nature (IWLN)—a phrase that avoids the term "Indigenous science"—interspersed with the regular science content. Non-Indigenous Ray

13 Cajete (2000)
14 Maryboy & Begay (2005)
15 Duran (2013)

Barnhardt and Yupiaq Oscar Kawagley: established the Network, developed polished lesson plans,[16] and meticulously conducted research with the students' standardized test scores that showed remarkable improvement.[17]

I initiated a two-year (1999–2000) research and development project (*Rekindling Traditions: Cross-Cultural Science & Technology Units*)[18] across Northern Saskatchewan (Northern Lights School Division) that mentored teachers in creating cross-cultural science teaching units, braiding together each way of understanding the universe. My agenda to advance the cross-cultural school science movement had been constructed from what I had learned from Indigenous educators. It was not a settler's agenda for deciding what is best. The topics were community-chosen so they related to students' everyday world. Elder Sanderson and I managed the project. An Elder in each location contributed ideas and perspectives to *Rekindling Traditions* units, which I wrote up under their supervision.

A major international project required three months of my time during a sabbatical leave in 2002. It became a rich encounter to my journey to be an ally. I visited Indigenous colleagues who had just finished a Japanese-sponsored international science education project concerned with the negative cultural impact of Western school science on Indigenous students' attachment to their own traditional culture. At three sites, coordinated research had been completed with:

1. Polynesian students in a rural town's school on the Big Island of Hawai´i;
2. Māori students in iwis, in Aotearoa New Zealand; and
3. Students from Aboriginal families in a western Australian cite.

By visiting the cites personally, I got to know: (a) what research they carried out, and (b) what the researchers concluded. For each cite, I wrote a report for the project's leader in Japan.

It became a stimulating crash course in learning diverse cultural ways of teaching Western science.[19] It gave me a hands-on experience comparing the nuances among different Indigenous groups. A special acknowledgment goes to Māori colleague, Dr. Liz McKinley, who did an excellent job at sensitizing me to the normal way English translations of Indigenous languages can

16 E.g., Kawagley (1995)
17 Barnhardt & Kawagley (2005)
18 Aikenhead (2002)
19 Aikenhead & Ogawa (2007)

misrepresent the authenticity of the Indigenous message. This phenomenon can cause an Indigenous person to replace their authentic Indigenous meaning with a Eurocentric meaning conveyed implicitly by the English meaning. This form of colonization was called "cognitive imperialism."[20] I acknowledge Liz's kindness and her relentless analysis of my English. She was a most effective mentor.

Back in Canada, my next step forward was to attempt to modify the province's science curricula[21] to include Indigenous perspectives; a political task more than an educational one.[22] An Indigenous sessional lecturer at the College of Education, Brenda Green (Dakota Nation) and I organized a small group to attempt the task.

Marie Battiste attended our first meeting to offer excellent political advice. We were aided by federal politics that included avoiding a lawsuit by Indigenous groups. It was settled by an apology from the Prime Minister for residential schools, and to a Truth and Reconciliation Commission to follow the apology. Next, Brenda was seconded to the Ministry of Education.

In 2008, Saskatchewan's Ministry of Education included Indigenous understandings in its revised science curricula,[23] helped by the collaboration with a representative group of about 10 Elders. The Ministry also decided to publish a new science textbook series (Grades 3–9)[24] that incorporated the Indigenous perspectives decided upon by that representative group of about 10 Elders. Each of the four units per book carried the summary of an interview with an Elder. I had the same two assignments by the publisher, Pearson Canada, for each book:

1. Assignment: To work with those Elders in composing English sentences that conveyed the Elders' authentic meanings, thereby avoiding "cognitive imperialism."[25] Thanks again to Liz McKinley for her mentorship.

   Over four years working on the textbooks, I learned about cross-cultural communication from what the Elders wrote for the textbooks, and even more so from their corrections of what I had written. I

20 Battiste (1986, p. 37)
21 aikenhead (1996, 1997, 2006b)
22 Aikenhead (2023)
23 Aikenhead & Elliott (2010)
24 Pearson Canada (2009–2012)
25 Battiste (1986, p. 37)

acknowledge their patience and hard work in teaching me some depths of coming to understand their culture. A University extension course in Plains Cree (Nehiyawewin), helped me became acquainted with its structure.

2. Assignment: To integrate their ideas into the text in detail beyond their interview summary. For example, I would add a relevant question to a list of questions the textbook authors posed; and I modified an English expression if it subliminally marginalized an Indigenous perspective.

Again, out of these discussions with the Elders came a wealth of in-depth learning on my part. This continued during four years.

Then Saskatoon Public Schools wanted a teacher professional development project written for teachers on how best to use the new textbooks. I was hired to assist Elder Darlene Speidel (Lakota) and her son "Waokiye" (Lakota for Knowledge Holder) Don Speidel. We spent the 2012–2013 school year cobbling together: (a) what the volunteer 12 teachers wanted to know; (b) what they experienced; and (c) their reaction to those experiences.

By observing Don Speidel convey an Indigenous way of living in Mother Earth with those teachers, I was able to detect the wisdom I had read or heard about.

My job included writing up the year's experiences, along with 14 co-authors, to capture their ideas, and some stories about their memorable classroom experiences as a result of the project.[26] It was published by the Saskatoon School Board as: "*Enhancing School Science with Indigenous Knowledge: What We Know from Teachers and Research.*"

An R&D project in Norway, "Local Cultures for Understanding Mathematics and Science" (LOCUMS), involved Sámi teachers and university science and math education professors during three week-long meetings spread out from 2010 to 2017. I was asked to be one of three international volunteer consultants. They were creating science and mathematics teaching materials in the native language. I am indebted to the Sámi Knowledge Holders for sharing their cultural ways of living. and for insisting that everyone have authentic experiences with deer, dog sled races, and ancient personal composing and singing called "yoik."

Coming to know in depth Indigenous ways of living in Mother Earth (both its understanding and its wisdom), naturally occurred when I and Indigenous

26 Aikenhead, Brokofsky, Clark, et al. (2014)

co-author and former graduate student, Dr. Herman Michell, wrote *Bridging Cultures: Scientific and Indigenous Ways of Knowing Nature.*[27] I acknowledge my privilege in learning so much from him.

A week-long national conference/culture immersion, "Indigenous Worldviews Informal Science Education" (I-WISE), organized by Dr. Nancy Maryboy, was held in Albuquerque N.M. in 2015. Its "Informal Science Education" refers to educational experiences in museums.

Elder Little Bear, his wife Amethyst First-Rider, and I were asked to lead one of six break-out groups that met almost everyday during the conference. The task also required a written report for the National Science Foundation, Washington DC. My draft of our report was scrutinized by Little Bear, First Rider, and Greg Cajete (another discussion group leader). Their feedback was a much-appreciated, in-depth, and further learning experiences for me. I acknowledge that useful feedback.

My last acknowledgment is to thank Elder Albert Scott (Nakawē) and Knowledge Holder Sharon Meyer (Nehiyaw). During the school year 2018–2019, Sharon and I mentored four rural math teachers (Grades 5–12) in designing and teaching two sets of Indigenous math lessons.[28] Albert and Sharon conducted a two-day cultural immersion, which contributed even more to my Indigenous coming-to-understand. I acknowledge their excellent work that produced a concrete recommendation to the Ministry of Education about how to include Indigenous mathematics, which is action-oriented, in the provincial math curriculum.

The above events summarized my story of becoming an ally of Indigenous peoples.

Eko'si (That is all I have to say.)

27 Aikenhead & Michell (2011)
28 Meyer, Aikenhead, Cardinal, et al. (2019)

ABBREVIATIONS/ACRONYMS

B.A.A.S.	British Association for the Advancement of Science
C.B.I.K.	Center for Braiding Indigenous Knowledges
C.C.I.	Canadian Climate Institute
ceo	Chief Executive Officer
C.E.P.A.	Canadian Environmental Protection Act
C.E.R.	Canada Energy Regulator
C.O.P.	Conference of the Parties
E.C.C.C.	Environment and Climate Change Canada
E.N.R.	Environment and Natural Resources
E.P.A.	Environmental Protection Agency
E.S.G.	Environmental Social and Governance
G7	The federal government ministers of industry and commerce from: Canada, France, Italy, Japan, the United Kingdom, the United States, and the European Union
G.C.	Government of Canada
G.F.A.N.Z.	Glasgow Financial Alliance for Net-Zero (U.N.)
I.E.A.	International Energy Agency
I.I.S.D.	International Institute for Sustainable Development
I.P.C.C.	Intergovernmental Panel on Climate Change

IWLN	Indigenous ways of living in nature (an authentic Indigenous translation of "science;" to *avoid* the outdated cognitive imperialist expression "Indigenous science").
I-WISE	Indigenous Worldviews in Informal Education
L.&D.	Loss and Damage
NDP	New Democratic Party
N.A.S.A.	National Aeronautics and Space Administration
NOS	nature of science
pc.	personal communication
R.&D.	research and development
SSI	Socio-Scientific Issues
STEM	science, technology, engineering and mathematics
T.R.C.	Truth and Reconciliation Commission
U.N.	United Nations

INTRODUCTION

By way of time travel, this book takes you back to Renaissance Europe (about 1400–1700 CE) just in time to witness when cultural values and beliefs about the solar system were about to change:

a. *From:* whatever the ancient Greeks believed. And their beliefs followed whatever philosopher Aristotle's pure mind told to him, for instance: "The Earth was at the centre of the universe," and therefore, so was everyone living on Earth. This made them feel special.
b. *To:* In about 1592, a university student by the name of Johannes Kepler did not find Aristotle all that believable. Johannes decided to gather evidence and rely on the evidence for the truth.[1] He reduced his evidence into groups of ideas (i.e., variables); a way of reasoning called "*reductionism*," or "reductionist thinking" used in doing experiments. It worked so well that over the years he came up with three laws of planetary motion. He firmly believed his third law represented: harmony as in harmony in the universe, harmony in humanity, and in his

1 Caspar (1993, pp. 43ff)

> spirituality. Back then, Kepler's work was called "natural philosophy." The label "science" would have to wait until 1831.[2]

From Kepler's Renaissance days, time travel will bring you past 1831, all the way to the 1980s when N.A.S.A. scientists were investigating a curiosity: Why was the Earth's atmosphere warming continuously year by year?

When the N.A.S.A. scientists made graphs of their data, they found that extra global heat generated in the atmosphere coincided with dramatic weather patterns at that time. They called this "*climate change*." Then when their graph line was *extrapolated* (extended further) into the future, it suggested that eventually deadly wildfires, destructive floods, and devastating droughts would most likely occur.

The predictions of such dangers from climate change became public news in the 1980s,[3] when N.A.S.A. scientist Dr. James Hansen testified at a Washington congressional hearing on the probable effects of climate change.

In 1988, Canada was the first country to organize and host an international conference on climate change. Three hundred delegates met in Toronto to discuss how to become sustainable in order to avoid the annual rise in the Earth's average global temperature, which happened since the beginning of the Industrial Revolution, in about the 1760s.

Taking seriously what he heard from scientists, Canada's acting Minister of the Environment declared that fossil fuel production must be cut back immediately, to reduce the concentration of greenhouse gases in the atmosphere.

That statement initiated a backlash from the petroleum industry, which promoted climate-change denial and lobbied governments not to listen to the environmentalists. This continues to this day with Big Oil's clever propaganda about their industry being sustainable, which masks their procrastinations that avoid significant action. Catherine McKenna (Former federal minister of environment and climate change) calls such talk, "climate-change cheap talk."[4]

The culture of Indigenous nations has always been sustainable, based on their spiritual relationship with Mother Earth, and their active and continuous participation in the Earth's evolution.

Geological, biological, or botanical occurrences interact among themselves and with instances of other events of scientific interest. *Holistic* reasoning is a superior way of coming-to-understand these broader interactions. Science

2 Orange (1981)

3 Suzuki (2023b, min 2.20)

4 Graney (2023a, b, c); McKenna (2023)

excels at reductionist reasoning; constructed from a reductionist perspective. *Both* a human-initiated (i.e., scientific) and a nature-initiated (Indigenous) ways of reasoning are needed to interact most effectively with nature to reduce the climate-change crisis we are in today.

Therefore, it is not surprising that at this moment, threatened by serious climate-change tipping points, that *collaborative* efforts among: (a) Western science, engineering, economics, politics etc.; (b) Indigenous ways of living in nature (IWLN); and (c) their developed wisdom-like perspectives; are just beginning to take action to eliminate the threat of today's crisis, with the hope that it is not too late.

In their everyday lives, citizens young and old have been confronted with contradictory scientific evidence, conclusions, and advice from the news media, friends and social media, concerning science-related issues. How do people decide: who to trust, what evidence to trust, and what sources to trust?[5][6][7]

In the prestigious journal *Science*, Osborne and Pimentel wrote: "We consider how scientists, science curricula, and science educators must help equip individuals to evaluate the credibility of scientific information, even if the science is beyond their understanding."[8] Three criteria were promoted: "credibility of its authors, the social practices of peer review, and the fact that it represents a consensus report of the relevant experts."[9]

Doubt and confusion arise when people lack a familiarity with the *nature* of science (NOS) itself.[10] For instance: What are some of the ways scientists develop dependable evidence or knowledge? What does "dependable" mean in the scientific community? How do scientists negotiate and decide what is a useful believable idea or observation?

Today, savvy citizens need to be aware of the limitations of science, and how science has evolved from being the "natural philosophy" subject of Kepler's (1571–1630) Renaissance times.

Other key NOS issues are: What ways has the scientific enterprise interacted with society today? and What interactive ways are quickly evolving today in response to the climate-change crisis? Scientists are dealing regularly with

5 Suldovsky et al. (2019); Cobern et al. (2022)
6 Aikenhead (2007); Lederman (2007); Duschl (2022)
7 Kolstø (2001)
8 Osborne & Pimental (2022, p. 246)
9 Osborne & Pimental (2022, p. 246)
10 Aikenhead (2007)

the media on issues of public interest. This was not how science *normally* operated. That is why science today is called "*post-normal science*."[11]

This nonfiction book, "*Confronting Climate Change*," begins by exploring Johannes Kepler,[12] a famous astronomer, mathematician, humanist, and notable refugee. His life exemplifies science being a human endeavor. This is another NOS idea. Kepler provides insights into today's scientific enterprise.

Between 1543 and 1687, the intellectual development describing and explaining our solar system flourished, because of Copernicus, Kepler, Galileo, and Newton. They were principally responsible for the Renaissance's "Scientific Revolution," as it is called today.[13] The culture of Europe changed accordingly.

Today, the main issue is how to protect the Earth and humanity from passing past a serious tipping point in *climate change*, from which there is no recovery. (Tipping points are described in detail in the book's subsection "Climate-Change Tipping Points," in Ch. 3.)

As reported so far in this Introduction, development in society mostly involved clever economists, inventive natural philosophers and engineers, as well as a few insightful politicians.[14] The Earth's protection against excessive climate change requires nothing less than a cultural revolution towards sustainability. One hundred years ago, such a disruptive revolution occurred with the transition from horse and buggy transportation to fossil fuel vehicles; initially expensive and inconvenient, but highly prosperous in the end. For humanity's survival, a disruptive Sustainability Revolution must occur.

A plan for it—the collaboration (between Indigenous leaders and Western scientists/engineers/economists/politicians) described in Chapters 5 and 6—is suddenly and unexpectedly taking shape internationally as this book goes to press.

The book has three goals for readers and their future everyday lives.

1. To understand the *nature of* science and scientists well enough[15] to challenge those people who: (a) misrepresent science and scientists; (b) dismiss legitimate scientists' professional research results; (c) demonstrate the "Dunning-Kruger Effect;"[16] (d) cannot distinguish between

11 Funtowicz & Ravetz (1993)
12 Caspar (1993)
13 Middleton (1963)
14 Gore (2007)
15 (Klopfer & Cooley, 1963)
16 Duignan (n.d.)

"the latest scientific knowledge," and "an outsider's interpretation;" and (e) engage in other tactics such as threatening violence.[17]

2. To recognize that Kepler's ideology[18] of *harmony* is strongly associated with our modern idea of sustainability. It holds the power with which to modify our culture into embracing both reductionist and holistic ways of reasoning, securely anchored within a sustainable context. The proposed sustainability revolution will be a *collaboration* involving climate-change reduction workers:
   a. Indigenous ways of knowing nature (a.k.a. IWLN); and
   b. Western science, economics, engineering, technology, enlightened politics, and litigation; along with;
   c. an accompanying shift from a profit-only society to a *reasonably profitable sustainable society*.[19]

This sustainability revolution will accelerate the elimination of the climate-change crisis that threatens humanity. Chapters 6 ends with a sound case for applying Chapters 1–5 to formulate a plan to eliminate the climate-change crisis by initiating a Sustainability Revolution inspired by the life work of astronomer Johannes Kepler, geneticist and conservationist David Suzuki, and economist Mark Carney; signifying harmony among the universe, humanity, and society.

The book's historical and scientific accuracy is expressed by referencing quotations and major ideas for readers to explore if they wish. This way, the reader chooses the topic, and this book does not become an encyclopedia.

17 Lomborg (2022)

18 In general, ideology refers to a doctrine of how one treats people.

19 In this book: (Mark Carney, Ch. 2); and (David Suzuki, Ch. 3)

· 1 ·

THE SCIENTIFIC REVOLUTION

The Natural Philosophers (Scientists)

Greek philosophy concerning the universe was rewritten during the European Renaissance (about 1300–1700 CE) by natural philosophers between 1543 and 1686, beginning with Copernicus' book and ending with books by Newton; listed in Table 1.1.

German natural philosopher Johannes Kepler (1571–1630) published many books that established evidence-based *descriptions* of the Sun located near the centre of the solar system, around which all the planets and stars revolved. Kepler created three laws describing those planetary motions in addition to many other discoveries and inventions ([12]). The result was Kepler's three *laws* concerning the solar system that Copernicus may not have recognized or even believed.

In understanding the nature of science (NOS), "scientific laws" tend to *describe* a phenomenon, while "scientific *theories*" tend to give a mechanistic (cause/effect) explanation. For example, the answer to the question: "What shape is Mar's orbit?" will be Kepler's first *law*–an ellipse. The answer to: "What makes

1 Caspar (1993)
2 N.A.S.A. (2017)

Table 1.1. The usual headliners for the first scientific revolution.

Copernicus	Kepler	Galileo	Newton
On the Revolutions of the Celestial Spheres. 1543	*The Optical Part of Astronomy.* 1604	*Galileo Against the People.* 1605	*Mathematical Principles of Natural Philosophy: The Motion of Bodies.* 1686
	The New Astronomy. 1609	*The Starry Messenger.* 1610	
	Harmony of the Worlds. 1618	*Dialogue Concerning the Two Chief World Systems.* 1632	*Mathematical Principles of Natural Philosophy and the System of the World.* 1686 "Principia" is short for "Mathematical Principles of Natural Philosophy"
	Compendium of the Copernican Astronomy. 1622		
	Rudolphine Tables. 1627	*Two New Science.* 1638	*Optics.* 1731

Mar's move in elliptical orbit?" will be an explanation drawing upon the concepts of momentum and gravity; inventions of Descartes and Newton, respectively.

A contemporary of Kepler's, the Italian natural philosopher Galileo Galilei (1564–1642)—the only famous scientist who people call by his first name. "Al's relativity theory" just doesn't sound right!

Thanks to the Dutch invention of the telescope, Galileo contributed new observations to astronomy, and also to some basic ideas in physics about objects in motion on the Earth. Here is another NOS idea: different technologies can give different kinds of observations, which may lead to changes in what scientists believe. That is one way how science progresses. You'll read of others as they happen. There is not just one "scientific method!"

Sir Isaac Newton (1643–1687), an English natural philosopher and theologian, had many insights about natural philosophy. Kepler's third law of planetary motion gave Newton the break-through hint that the sun pulls on all planets in a certain way. He used some advanced geometry to figure out the details for universal gravitation based on Kepler's data. It became a very controversial idea at the time—a huge force acting on the moon, the planets, and causing the ocean tides, without touching them. French scientists treated it as a joke.

Another scientific method happened here: one natural philosopher's idea was picked up and used in a different context. The story about an apple following from a tree was a silly answer to the French. His advanced geometric derivative of his gravitation law was published in his 1686 two-volume book known as *The Principia*; shorthand for *Mathematical Principles of Natural Philosophy:* (1) *The Motion of Bodies*, and (2) *The System of the World.*

Newton made it very clear that his insights relied on the conclusions from Kepler and Galileo. During his life time, Newton had an increasing number of international colleagues who investigated many topics in natural philosophy.

Newton is also known for describing general beliefs and strategies that he used to come up with his new knowledge (i.e., a primitive version of beliefs and strategies associated with what we call "the nature of science;" NOS). Unfortunately, others in the early twentieth century misunderstood what he wrote. They oversimplified it by the phrase "the scientific method."[3] The general public today who prefer oversimplifications in their daily life, tend to use the phrase "the scientific method" in their simplistic descriptions of what scientists do. Correcting this over-simplification is on the list of NOS ideas. Today, most science deniers rely on this misleading over-simplification to support their anti-science positions. For instance, "She didn't follow the scientific method, so who is going to believe her?" But the generality (i.e., the scientific method) is make-believe.[4]

The period 1543–1686 changed natural philosophy in so many basic ways that today we refer to the period as "The Scientific Revolution."[5]

Fundamental Changes to the Culture and Language of Europe

Not only did Johannes Kepler provide us with the evidence-based truth about the solar system, he identified modern values useful to both the progress of natural philosophy and, unexpectedly, a "roadmap" with which to tame climate change today. Under his influence he helped alter a foundational understanding in his culture's epistemology–*how* do we know what we know?

Kepler was taught the renaissance version of a more specific question: How do we know if something is true or not? The ancient Greek scholars' cultural

3 Rudolph (2005, p. 341)
4 Rudolph (2005, p. 368)
5 Cohen (1960, p. 63, 1971, pc)

belief was clear. With the Earth at the centre; the moon revolves around it *in a circle* at a *uniform speed*; as do the Sun and five other planets known at the time. Lastly, the outer reaches of the universe had an outer hollow sphere that holds the stars in place.

The moon, Sun, and each planet were attached to its own hollow, crystalline sphere; with the Earth at the centre. Think of the crystalline spheres as being nested with a space between each sphere. Cut an onion in half and notice the exposed flat centre face. That is the structure held by the ancient Greeks. *Is it true?*

Kepler contributed significantly to the changes in the scholarly Renaissance culture; which:

1. *Changed from* the ancient Greek's belief that truth arises out of Aristotle's or Plato's purity-of-the-mind;[6] *to* our modern evidence-based reductionist thinking that has endured in science for almost four millennia. "Thus, the geocentric theory of the universe is the logical consequence of [Aristotle's] doctrines of natural motion and natural place. The universe was made to fit the doctrines, not the doctrines to fit the universe."[7] Kepler's contribution was to establish the value that the data (i.e., observations of the universe) must fit any proposed theory or law and become the era's doctrines in natural philosophy.
2. *Changed from* the belief that humans were located at the centre of the universe, *to* the realization they merely inhabited the third planet in the solar system. However, the "I-am-the-centre-of-the-universe" ideology has not yet been extinguished from most cultures, it would appear, judged by the behavior of today's climate change deniers.

N.A.S.A. attributed many "firsts" to Kepler's list of accomplishments in Natural Philosophy;[8] for instance:

- First to explain that the tides are caused by the Moon.
- First to establish the truth of a *heliostatic* solar system (a stationary sun), in his book series: "Compendium of the Copernican Astronomy," published in 1618, 1621, and 1622; where "compendium" means "textbook." Science historian, I. B. Cohen, concluded, "It may readily be seen that

6 Linnebo (2018)
7 Middleton (1963, p. 23)
8 N.A.S.A. (1917)

the publication of Copernicus' book in 1543 did not of itself achieve a revolution in physical or astronomical thought."[9]

Kepler did foresee the need for our modern way by which scientists share their observations and ideas. In his time, most natural philosophers, such as Tycho Brahe and Galileo, were reluctant to share what they had discovered for fear that it would be stolen by another natural philosopher who would take credit for the observation or idea themselves.[10]

Kepler's wish became a reality when a politically sanctioned group of natural philosophers in England established the Royal Society in 1660. Its purpose was:[11]

> to bring together leading scientific minds of the day, and to become an international network for practical and philosophical investigation of the physical world. This indicates the beginning of natural philosophy's professional status. At first, groups of physicians and natural philosophers held meetings at a variety of locations, including Gresham College in London, which became The Royal Society's home.

As significant as this new organization was, they retained the same name, "natural philosophy." It took another 170 years before it was changed to "science." This happened when a new image of natural philosophy was required in the nineteenth century. Its public image needed to become professionalized.[12] Very purposefully and politically, the name *science* was chosen to replace *natural philosophy* during the birth of a new organization in 1831, the British Association for the Advancement of Science (BAAS). To garner public support, they wanted to create the image of "a group of men united by a common dedication to the investigation of nature."[13]

At the time, "science" meant any knowledge generally. But the new meaning was added to the English lexicon.[14] However, it took half a century for BAAS members to accept grudgingly the term "scientist" (i.e., cultivators of the new science) as a replacement for "men of science."[15]

9 Cohen (1960, p. 61)
10 Caspar (1993, pp. 120–121)
11 Science Museum, London (2019, Aug 7, website quote)
12 Layton (1986)
13 Yeo (1981, p. 69)
14 Orange (1981)
15 MacLeod (1981)

Harmony

During his life, Kepler emphasized one value over all others: *harmony*—his key message to the world.[16] He expressed that feeling about his mathematical third law of planetary motion because each planet's motion could be described by one mathematical algorithm. The algorithm exemplifies "the big picture," a characterization of holistic reasoning.

Today when asked, "How do you know that?" an insufficient answer is, "That's what Aristotle wrote." The answer requires either an evidence-based response (reductionist-like reasoning), or a plausible description of how it is related to a fundamental "big picture" (holistic-like reasoning). For Johannes Kepler, his big picture was often harmony.

He explained the title of his book, *Harmony of the World*, to his wife, Susanna. It referred to the harmony reached among: the *world of nature, the world of humankind*, and the *world of their spiritualities*.[17] Harmony was also expressed, for instance, through the harmony of the music emanating from the motion of all the crystalline spheres that comprised the universe.

On the one hand, he and his family were tossed about by the disharmony among different religions, including within his own Lutheran faith. On the other hand, Kepler achieved a high degree of *harmony* through his mathematics and astronomy. But unfortunately, he did not live long enough to witness the harmony brought about by the 1648 Peace of Westphalia that ended the 30-Years War; nor the harmony achieved within natural philosophy in 1660 when the Royal Society was inaugurated in London.

There will be another opportunity (from 2024 to 2050) to achieve a significant degree of *harmony* among: (a) natural scientists, engineers, mathematicians, and social scientists—those who advocate green investments, and those who achieve the facility to add holistic thinking in the context of sustainability to their repertoire of reductionist reasoning; and (b) a general public who are sufficiently savvy about the nature of science and scientists (NOS), to become politically supportive in solving the climate change crisis. One key NOS item for being supportive is the nature of scientific measurements; it is a crucial topic.

16 Caspar (1993, pp. 265–269)
17 Tiner (1977, p. 173)

A Measure of Measurements

Kepler highly valued the Danish Prince Tycho Brahe's precise observations of the planet's location in the night sky over many years. In fact, they were valued so greatly that in Kepler's mind they became *absolutely* accurate and *objective*. However, David Fabricius, a well-respected German astronomer and friend of Kepler's from another university, believed that *no measurement is absolutely accurate and objective.*[18] Instead, he thought natural philosophers do their best when they simply *reduce* a measurement's subjectivity so the measurement becomes *accurate enough*. Here is how the issue was resolved for Kepler.

One day, Fabricius visited Kepler at a time Kepler was dealing with a measurement issue. Kepler had discovered a difference of 8 minutes of arc (an angular measurement unit) between an observation by Tycho Brahe and a point on Kepler's drawing of Mar's orbit. This measure of 8 minutes needs to be put in perspective. There are 21,600 minutes of arc around a complete circle.

When Fabricius said that 8 minutes is too small to worry about, he was expressing a scientific value, different from the one Kepler had learned at Tübingen University. To solve the 8 minutes of arc problem, he had to set aside what he was taught and embrace a new idea. For any one measurement, there is a natural legitimate *range of observable results* because of something being measured by different people and/or by different instruments. The "real" result lies within this range. If 8 minutes of arc lies within that range of a legitimate results, then it becomes equivalent to any figure within that range. That was why Fabricius did not worry about such a small difference. He believed there was no *significant* scientific difference between Brahe's and Kepler's measurements.

The following is an equivalent everyday experience by a parent today when a curious mother asks her teenage son, "What time did you get home last night?" Her son answered, "Eleven forty-five, *more or less*." We might wonder what "more or less" means exactly. We might even wonder if son and mother had the same figure in mind.

If they did, the everyday language of "more or less" means the same thing as the more precise "small range of legitimate observable results" in science. This range is called the "*margin of error*," or "*confidence interval*," or unofficially "the spread." The narrower it is, the more *confident* you are that your measurement is more accurate.

18 Tiner (1977, p. 130)

The take away from this example is straight forward. There is *always* a margin of error, even with the most advanced technology! This is just one of many "*nature of science*" (NOS) ideas for understanding something deeply. You find those ideas by looking, listening, or reading deeply into the scientific culture.

For instance, most court cases about driving under the influence of alcohol are usually about the confidence interval of a person's breath-measurement. In many countries, the legal margin of error of ± 0.01 ("plus or minus 0.01" on a scale of 0.00 to about 0.30).

For instance, an accused driver blew into a tube connected to a breathalyzer. It showed a reading of 0.09. So the real scientific reading, expressed with its confidence interval, was actually "0.09 ± 0.01." This meant that we can be confident that the correct breathalyzer reading was somewhere from 0.10 to 0.08. The driver's lawyer pointed out to the judge that the reading could legitimately have been 0.08. And this is a legal value. The judge agreed. The accused was found not guilty.

Here is a more familiar scientific example. With the human eye, few people can read a typical school protractor as precisely as ± 0.01°. Its reasonable margin of error is reasonably about ± 1° *for each angle measured.*

Suppose you are confident that a protractor's measurement is somewhere "between 45° to 46°." By convention, this *includes* both 45.0 and 46.0. That is your confidence interval written in English. To translate this English into a scientific measurement, first find the halfway point between the two extremes of 45° and 46°. The halfway point here is 45.5°. Then your scientific measurement with its confidence interval is expressed as: "45.5° ± 0.5°."

Now consider two students' measurements. Student A's measurement was 45.3°, and Student B thought it was 45.7°. Here is the key question to pose: Are the two observations *significantly* different? That is the question Kepler posed when he thought his measurement of 8 arcs was significantly different from Brahe's measurement of a planet's position on a sky chart. But Fabricius advised Kepler to forget about the 8 arcs of difference, and think of them as more or less the same measurement.

Next, you are about to repeat what Fabricius likely worked out in his head. Then you can follow the same procedure in the future with a pencil and paper yourself.

Students A and B said their measurements were 45.3° and 45.7°. But those were incomplete. They lacked their confidence intervals. So here is the way to construct them. First make an educated guess on how accurately you could repeat a measurement and be confident that they will always be *within the same range* of observations. Angle measurements with a cheap student projector are likely to land within a range of 1.0°. That is a total spread that you expect to be *within*. So you pick the *centre* of the spread of 1° and use the "± 0.5°" to represent the spread of 1°. This makes student A's measurement 45.3° ± 0.5°, and student B's 45.7° ± 0.5°.

This is the same as: |44.8° to 5.8°| and |45.2° to 46.2°|.

Now compare the two different measurement *intervals* mathematically or visually like this:

Student A's confidence interval is: |44.8° to 45.8°|
| *overlap* |
Student B's confidence interval is: |45.2° to 46.2°|

As indicated by the *overlap* information, just above, the two confidence *intervals overlap* between 45.2° and 45.8°.

Here is the easy part. Do they overlap or not? Because if they do overlap, as they do, there is a chance they could be the same result. In statistics, this means that the two measurements are ***not*** *significantly different* (a double negative). In science, they are treated as equivalent measurements. That is how the measure of measurements works in science. It is also a point at which Kepler put aside his ancient Greek idea that measurement numbers are absolute points, rather than a spread possible points.

What if Students A and B each had thought their measurements were more accurate than first assumed. Consequently, they assigned a confidence interval of 0.5°, which amounts to ± 0.25°. You do the math yourself. You should discover that both confidence intervals still overlap, although less so. The two are still *not* significantly different. You can treat them as equivalent measures.

Here is one simple fact about confidence intervals you should know. Suppose you were working as an architect's helper. You had to measure three different angles printed on a blueprint design for a kitchen remodeling, and then add them up. This requires three different measurements. Assume a confidence interval that covers one whole degree (i.e., ± 0.5). For the sum of the 3 blueprint angles measured, logically speaking, the confidence interval of the sum is ± 1.5°. (The margins of error are added whenever you add *or subtract* measurements.) If your sum is 181.0°, then you are confident that the three measured angles added together could lie from 179.5° to 182.5°. You would report your sum as: 181.0° ± 1.5°.

For the record, another strategy goes with *multiplying or dividing* with measurements expressed with confidence intervals. The rule is the same for both multiplying and dividing.

Rule: Change each ± value into a percentage of the measurement. Then add the percentages. Finally, convert that percentage of the calculated measurement back into a ± value.

As mentioned above, Kepler led the Renaissance's natural philosophers away from Ancient Greek thinking and into modern ways of thinking that guide science today. Fabricius' advice freed Kepler to continue to develop his third law of planetary motion. This situation reveals the following nature of science (NOS) topics:

1. Science progresses through contacts with colleagues (e.g., by attending conferences, or visiting other universities or industries); and
2. Instead of getting an *objective* measurement (i.e., meaning that no uncertainties or biases are associated with it), scientists *reduce its subjectivity* as much as humanly possible, in keeping with the situation at hand.

The Negativity of Change

Three astronomers (Aristarchus of Samos, Copernicus, and Kepler) promoted models of the universe, for which humans were not located at the centre.

Some Renaissance people in positions of power were so upset over such an anti-Aristotle message that these "radical" believers in a different idea of the universe were punished, sometimes burned at the stake if they did not change their belief. Some empowers or dukes declared war on states that dared to convert to the new way of thinking.

Similarly, the general public did not tolerate change very well, especially those who were convinced they were indeed the centre of the universe. They held superstitions about sudden changes in the night sky (e.g., the appearance of a comet meant an upcoming death of an emperor). As well, they usually did not tolerate people who lived a life quite different from theirs. Retaliation might include accusing such odd people of witchcraft.

Along with the Renaissance's motivation for progress in natural philosophy, there were new demands that came indirectly from the public. In other words, there was a two-way interaction between natural philosophy and

Renaissance society. For example; astrology was an important feature of their culture. Therefore, Tübingen University taught it along with mathematics and astronomy. This triad was customary in those days; part of the culture.

This interaction between science and society in our Digital era has become even stronger than ever. Interacting includes solving pandemics and attempting to avoid a climate-change crisis. We were promised that science and technology would somehow help solve any societal problem. But those decisions about science helping society, always involve:

1. peoples' personal values, prejudices, and participation in establishing how to put the science into public practice; and
2. scientific values that guided the development of science in the first place.

On a practical note, some of Kepler's important decisions were about his sources of income. He always had to worry about his finances. Some money came from teaching. That was before his Linz Protestant church banished him for his humanitarian way of treating others equitably.[19] Then his income came from individual patrons, such as the Duke of Württemberg, Emperor Rudolph II, and Emperor Matthias. But getting the gold coins they promised him was often a challenge or a failure.

Natural philosophy (science), math, and technology interacted with the culture of Renaissance Europe to try to ensure a stable economy. And today, our Digital Revolution era promised to improve society's economy and the standard of living for citizens, through advances in science, technology, engineering, and math.

This brings us to the need to know something about economics, a major influence on science today as science interacts with society; for instance, at the UN's yearly climate change meetings called "Conferences of the Parties" (COP). This book points out that a sustainability revolution involving both science and society equally is needed to solve our climate change crisis. The first revolution's successes give us guidance for proposing a second revolution today. Chapter 1 has laid the groundwork for what follows.

19 Caspar (1993, pp. 213–220); Tiner (1977; pp. 165–167)

· 2 ·

THE WORLD HAS MOVED ON

Introduction

From the scientific revolution, the world has indeed moved on. Here we follow its trail to the present time by noting how the social science of economics has changed over the years. Four time periods are visited: (a) the Renaissance (approximately 1500–1700); (b) the Industrial Revolution (approximately 1760–1840), the root cause of climate change; (c) today's Digital Revolution (approximately the 1960s and on); and (d) the climate-change-crisis challenge that we have entered very recently.

The World of Economics

Economics is one branch of what today we call the "social sciences," which include: sociology, psychology, architecture, education, etc. Thus, economists are social scientists.

In very general terms, the "natural sciences," as we call them today, include: biology, chemistry, physics, mathematics, statistics, engineering; etc. These different academic subjects have their own way of measuring or assessing what they investigate. Economics has been important in its interaction with the natural sciences.

The Renaissance's culture transition *from* Ancient Greek thinking *to* modern ways of thinking about the universe foreshadows a current economic transition from (a) today's modern ways of thinking, to (b) survival ways of thinking and acting in response to the climate change crisis. This transition is best understood by the series of changes that economists have lived through, from the Renaissance era today's climate-change-crisis era. I begin with an economist's point of view.

Economists

Like other expert scientists, economists must: (a) earn a Masters or PhD degree in economics, and (b) choose an area of specialization for their research. They study how people individually and collectively distribute scarce resources for businesses, industries, and consumers.[1]

Economists also identify potential trends from which to make economic forecasts, based on their *measurements* of a country's economic "indicators." These indicators include: the gross domestic product, retail sales, industrial production, employment and unemployment data, the consumer price index, and public surveys regarding consumer confidence. In other words, economists investigate societies' financial resources and the production of goods and services, measured by various indicators.

For instance, if there were a change in spending trends in a particular industry, its administrators may want to hire an economist to do some research on: (a) what will be the next popular product? and (b) what are the forecasts for job market growth in that industry? Each year, a Nobel Prize in Economics is awarded to the most outstanding economist or team of economists.

Economists measure such indicators as: poverty levels, proportion of citizens in hunger, quality of education, sustainable communities, wealth of some individuals, success of industrial domains and their meeting gender equity.[2]

World-renowned economist, Canadian Dr. Mark Carney, was born in 1965 at Fort Smith, Northwest Territories. His family moved to Edmonton, Alberta, where he attended school. Eventually he earned a PhD in economics at Oxford University. Between 2008 and 2020, he served as a Governor of the Bank of

1 Kenton (2022); Hayes (2022)

2 Kleespies & Dierkes (2022)

Canada, and then Governor of the Bank of England. He supports a sustainable approach to economics.

During his studies of economics, Carney detected there were three different economic periods of history from the Renaissance up to current times: the Values Society, Profit Society, and Sustainable Society.[34]

Values Society

In Renaissance times, political leaders (e.g., emperors, kings, dukes. etc.) held *values* that guided the decisions they made for society (e.g., Will there be peace or war? or What major construction projects will we start?). For this reason, Carney labelled Kepler's society a "values society." Decisions were made according to the ruler's values.

Carney is: (a) the Special Envoy of the UN Secretary General for Climate Action; (b) UK Prime Minister's Advisor for Climate Change; (c) chairman of the Committee on the Global Financial System at the Bank for International Settlements; and (d) chairman of the Financial Stability Board, based in Switzerland.[5] Simply put, he excels at being an economist, an academic, and a humanitarian. He certainly is a potential Nobel Prize winner in Economics.

Renaissance Europe's economic system believed that peoples' wealth was both limited and unchangeable. People who prospered were those who hoarded their gold coins or were able to place a tax on items imported from other states or countries. This type of economic system is known as a mercantile system.[6]

Profit Society

Beginning in the late eighteenth century, European economic systems began to change radically. This change from the mercantile system was due to a Scottish economist, Adam Smith, who published a famous book, *The Wealth of Nations* (1776). It eventually caused nations to replace their values society with a "profit society."[7] Peoples' actions were justified and assessed, based on how much profit was gained.[8] The best decisions were those that led to the

3 Carney (2020a, 2002b, 2021)
4 Carney (2020c, Utube version)
5 Carney (2021)
6 Carney (2021)
7 Carney (2020a)
8 Stout (2002)

greatest financial profit. Natural philosophy and mathematics became enablers to a profit society.

Business and industry profits often rose at the expense of depleting the Earth's resources and its biodiversity. At the same time, both industrial and consumer world-wide pollution (e.g., related to petroleum powered vehicles) increased to such an extent as to cause the devastating climate change phenomenon. Carney[9] pointed out that the "free-market" (i.e., "open-market") ideology of pure profit societies ignored the cost of depleted resources and increased pollution.

By tradition, such costs did not belong in an economist's bank ledger or a mathematician's algorithm. That would go against a free-market ideology—pay attention only to business profits and loses; ignore the negative consequences that consumers must deal with, such as climate change.

Without noticing it, subtly a very *major* shift in peoples' ways of valuing occurred. Science and mathematics helped to influence how the general public valued things. Scientific and mathematical ways of understanding are highly analytical; they *reduce* an issue or observation into identifiable parts, and then isolate each part by working on one separately (isolated) from the other parts. As mentioned in Chapter 1, this approach is called "*reductionist thinking*." It is the bread-and-butter of scientific reasoning and experimentation.

For example, Einstein *reduced* the whole universe into two categories: mass and energy. He then cleverly found a way to relate the two: "Energy equals mass times a constant number squared:" $E = mc^2$. The constant number "c^2" is the speed of light multiplied by itself.

Reductionism is not necessarily bad, but it can be when reductionist science ignores its influence on society, or chooses to isolate itself from society. As a result, the topic of ethical versus non-ethical behavior is often set aside and isolated by natural scientists, economic scientists, politicians, and some of the general public. They focus only on their financial gains, and ignore how those gains were acquired and at what cost to others? This major ethical issue is given low priority by the culture of a profit society.

Accordingly, the economists' ledgers have separated and ignored the consequential environmental and human costs to profit making (see examples in the subsection "Climate Change Damage in 2022" below). But some economists justify themselves by claiming they are thinking scientifically. This is a subtle way science and mathematics can be enablers to the profit society and its

9 Carney (2020b; 2021, p. 139)

open-market ideology. They arbitrarily hide ethical considerations from public scrutiny by claiming a Platonist purity sense of "objectivity;" in other words, "an opiate of the academic."[10]

The opposite of reductionist thinking is *holistic thinking*, in which a specific topic of interest has meaning in terms of its relationship with a much bigger (whole) picture or concept.[11]

Free-market economists measure almost everything according to their reductionist economics. At the same time, people tend to ignore hard-to-measure, yet important outcomes. Until recently, this narrow point of view controlled what got measured.

For instance, economists used to state, "The cost of oil is $50 a barrel." But now it is becoming more like, "The cost of oil is $50 a barrel plus the rising costs attributed to climate change,"[12] such as devastating wild fires; as well as: coastal land disappearance, brutal wild storms, deadly mudslides, massive floods, destructive droughts, and forced refugee migration.

In reality, these consequences are due in part to a profit society's ideology[13] (i.e., the free-market is always right; go ahead and make a maximum profit possible; and don't worry about the consequences to others). This ideology is well known to both disaster victims and international investment brokers. It also reinforces the "I'm the centre of the universe" ideology.

However, some economists are now beginning to create change in their cost-benefit analyses by including the cost of issues related to losses by people and the environment.[14] There is a real cost to climate change, rather than some fictitious free-market theoretical rationalization; such as "it's the cost of having a vibrant economy."

Sustainable Society

Climate change crisis talk emphasizes 1.5°C and 2.0°C as found in the quote just below. The two temperatures are discussed later in Chapter 3. Mark Carney and other research economists pointed out that a new economic transformation is already underway: a move from a profit society to a *sustainable*

10 Aikenhead (2008)
11 Lux (2021)
12 Hendry (2014, p. 63)
13 In general, ideology refers to a doctrine about how a person or institution should treat people.
14 E.g., Carney (2021); Stout (2002)

society. You cannot make a profit if there is no habitable planet. Reis[15] characterized the crisis as a matter of raw *human survival.* The Secretary-General of the UN, Guterres, quoted figures that tell us that a serious tipping point is dangerously near.[16]

> A new flagship UN report on climate change out Monday indicating that harmful carbon emissions from 2010–2019 have never been higher in human history, is proof that the world is on a "fast track" to disaster, António Guterres (UN Secretary-General) has warned, with scientists arguing that it's 'now or never' to limit global warming to 1.5°C. The UN chief added: "This is not fiction or exaggeration. It is what science tells us will result from our current energy policies. *We are on a pathway to global warming of more than double the 1.5°C goal.*"

Carney said, "The major international investments are turning nations away from a profit society towards a *sustainable society*, in which some profits are being made, nevertheless."[17]

Uncooperative businesses, industries, and organizations are now being called out in published reports, with headlines such as: "Oilsands putting profits ahead of lower emissions, report says,"[18] referring to: "Canadian oil sands companies have done little to follow through on their public pledges to reduce their greenhouse gas emissions, despite raking in historic profits in 2022."

Moreover, "Under pressure to resign for declining to accept the scientific consensus on global warming, World Bank President David Malpass said on Thursday it was clear greenhouse emissions are causing climate change and defended his record as bank chief."[19] Public pressure can sometimes be an instrument of choice to make things happen.

15 Reis (2022)
16 Guterres (2022, website quote)
17 Carney (2020b, transcript quote)
18 Graney (2022, website quote)
19 CBC News (2022, website quote)

· 3 ·

SOCIETAL DECISION MAKING AND POST-NORMAL SCIENTISTS

A recent post-normal science scenario from the COVID-19 pandemic will illustrate:

1. Just how badly the rift has become between scientists and some of the general public; and therefore, by implication
2. How badly secondary school science needs to emphasize teaching the *nature* of science, including science's interactions with society[1] for the large majority of high school graduates (85%)[2] who see themselves doing well outside of science-related fields of employment.

The consequences of some science-related societal issues have grown extraordinarily serious internationally. The COVID-19 pandemic was, and continues to be, fatal for millions of citizens. Moreover, deaths and horrendous *debts* are already mounting from the current devastation of climate change.[3] Carney[4] explains that the values that guide such decisions about what to do about climate change have been invariably political.

1 Duschl (2022); Klopfer & Aikenhead (2022)
2 Card & Payne (2017); Frederick (1991)
3 James et al. (2022); Reguly (2022a)
4 Carney (2021, p. 594)

Some scientists have already incorporated "positive politics into their new post-normal science,"[5] but some have faced very negative political tactics from certain segments of society.

For example, Dr. Anthony Fauci, director of the U.S. National Institute of Allergy & Infectious Diseases (NIAID), was interviewed by Nahlah Ayed[6] during a Canadian Broadcasting Corporation's (CBC) public radio broadcasting program called "Ideas."

AYED: In the [nineteenth] century in the West, science displaced religion as an ultimate authority in society. But now the authority of science in the public sphere is in danger of being displaced by social media and political rhetoric. I'm wondering what you think happened?

FAUCI: This is something that is so disturbing to me as a physician, as a scientist and as a public health person. If ever you could imagine the worst possible environment into which a global pandemic emerges, it would be in an environment of anti-science [and] complete normalization of lies.

It is just mind-boggling. It's when you have this kind of combination of divisiveness with the complete accessibility and spread of complete falsehood. . . .As a member of society, reaping all the benefits of being a member of society, you have a responsibility to society.

AYED: I'm wondering if there's anything we can learn from eradicating viruses that could help us in eradicating misinformation?

FAUCI: I lose as much sleep . . .worrying about the overall implications of the rampant spread of misinformation and disinformation on society. . . . It's as easy to say something crazy as it is to say something that's based on years of scientific evidence. . . .[And then comes a] false equivalency. "You know, a Nobel laureate who discovered this says this." "But Joe Johns on their Facebook said that." That has to be frightening, because it's happening.

AYED: How do we restore scientific authority in the public sphere?

FAUCI: One of the things that has been the source of considerable confusion is the lack of appreciation that science is evidence and data. . . .Science evolves and science corrects itself. There are many people who interpret that as flip-flopping.

In Dr. Peter Hotez's book, *The Deadly Rise of Anti-Science*, he discusses how an anti-vaccine movement became a dangerous political campaign.[7] He stated, "more than 200,000 unvaccinated Americans needlessly died."[8] This death toll far exceeded that of the twin towers attack by a different dangerous political

5 Funtowicz & Ravetz (1993, p. 739)

6 Ayed (2021, transcript quotes)

7 Hotez (2023)

8 As quoted in Amazon.ca (n.d.)

campaign. Those in the anti-vaccine campaign assume the role of a medical theorist camouflaged by the freedom of speech mantle.

The Nature of Science

This sorry state of affairs suggests that a significant number of science deniers among the general public are missing the fundamentals of the nature of science (NOS) needed to be savvy enough to participate reasonably in a democratic society, rather than participate with brute force, as happened with Dr. Fauci. Some claim that science does not tell the truth, in an absolutist Plato-like sense; because science seems to change what it claims to be true. The NOS content they are missing is *science is not the Plato-like (Platonist) truth*! When scientists change what they claimed earlier, they did not lie to you, they learned more in the meantime. The public needs to understand the NOS idea: *Science produces tentative but trustworthy evidence and knowledge.*

These anti-science people are using "truth" in an ancient Greek Platonist way: whatever comes out of Plato's or Aristotle's purity-of-mind, described in Chapter 1. These absolutist-truth seekers were out of date by the late-Renaissance natural philosophers' standards; and certainly irrational in the context of today's Digital Revolution standards. But in the context of persuasion on social media, the science deniers appear to be exercising their right to believe as if they are the centre of the universe.

The above crucial NOS concept tends to get ignored in science curricula that emphasize pre-professional science training for *all* students.[9] Unfortunately in that milieu, high school graduates[10] tend to come to the erroneous conclusion that scientists "always pretend to have the whole truth."[11]

The highly respected physicist John Ziman anticipated this public pushback from misinformed citizens: "I interpret it as symptomatic of the transformation of science into a new type of social institution. As [scientific ideas] become more tightly woven into the *social fabric* of science, scientists are having to perform new roles."[12] How does society prepare future scientists for new societal roles in "*post-normal science?*" This is a very different new era for most scientists to be in. It is strongly run by the interaction of science and society.

9 Aikenhead (2006a)
10 Aikenhead (2023, pp. 260–261)
11 Zeyer (2023)
12 Ziman (1998, p. 1813)

1. To update the small majority of high school graduates (about 15%)[13] who are academically prepared to pursue university natural science/engineering programs, how will their studies include connections to sustainability, for instance? How will they interact with the public in post-normal science?
2. Equally important is the vast majority (about 85%) of high school graduates who will become tomorrow's general public participating in post-normal science. If they simply memorize a catalogue of NOS concepts, their knowledge will most likely be superficial and therefore, mostly unusable out in the real world.[14]

Both: (a) an in-depth understanding of NOS; and (b) a familiarity with science's interactions with society; tend to help high school graduates accomplish the following:

- Reduce their susceptibility to misinformation and conspiracy theories about science.[15] "At this point in the pandemic, it's misinformation that's killing people."[16]
- Figure out who and whose data to trust;[17]
- Develop the competence to challenge irrational and the misinformed people who they encounter;[18] and
- Remember details that show that science is a human endeavour.[19]

When confronted with a public disagreement over a science-related societal concern such as global warming, consider organizing your thoughts or discussions around five points:[20]

1. What are the facts? How were these facts produced? What is their degree of uncertainty?
2. What is valued and what is devalued?
3. What is at stake? Who benefits from various versions of the facts?

13 Frederick (1991); Card & Payne (2017)
14 Bromme et al. (2015); Cobern et al. (2022, online)
15 Duschl (2022)
16 Bigham (2021)
17 Hendriks et al. (2020); Suldovsky et al. (2019); Osborne & Pimental (2022)
18 Kolstø (2001)
19 Zeidler, Walker, Ackett, et al. (2002)
20 Barwell (2013, p. 12)

4. Who benefits from emphasizing some values and not others?
5. What decisions are possible? What decisions are not possible? What action can we take?

Our sustainability crisis requires dedicated, post-normal and sustainability-minded: scientists, engineers, architects, politicians, artificial-intelligence mathematicians, educators, and high school graduates. This is in addition to: dedicated economists, politicians and community leaders who can help transform the mainstream culture of society, from a profit-only society to a reasonable profit sustainable society.

Together, these people need to create ideas to enhance sustainability in every way imaginable. This would be similar to the scientific revolution that replaced Greek values and ideologies with modern values and ideologies. The outcome was an authentic understanding of the universe and how it worked. Today, the outcome concerns *humanity's* survival. The stakes have never been higher.[21] So far, attempts to achieve this have been to organize annual international meetings.

Key International Meetings on Climate Change

The very first meeting on climate change occurred over a century ago among a small group of research scientists working in Aotearoa New Zealand. They published their results in a 1912 article, which was apparently ignored and forgotten. They had concluded that burning coal will eventually cause extraordinary weather consequences for humans. Back then, two trillion tons of coal were burned globally during one year, thereby sending seven trillion tons of carbon dioxide into the atmosphere where it acted like a blanket to heat the Earth. The authors warned that the consequences for the Earth's weather would become extraordinary in several centuries.

The original article is under strict copyright protection today. The newspaper *USA Today* reported that it was authentic.[22] Copies surfaced in the 1980s after climate change became a public topic. Today the original 1912 article apparently appears on some social media.

21 Carney (2021)
22 Sadeghi (2023)

Theoretical physicist, Edward Teller, senior scientist on the U.S. hydrogen bomb project, did publicly warn the American Petroleum Institute in 1959 that burning fossil fuels produces greenhouse gasses which warm the planet. And at the present rate of the atmosphere taking in those gasses, by the year 2,000 the results could be catastrophic.[23] The message was not news worthy at the time, but the industry cannot claim they did not know. This information may count as legal evidence in law suits against Big Oil[24] (a nickname for the major Oil companies) for the damages caused by the product hey sold, making outstanding profits (Chapter 6).

The climate change issue had disappeared publicly until 1988 when research on it by Columbia University's climatologist and astronomer, Dr. James Hansen,[25] became public information during his U.S. Congressional testimony on climate change. Later in life, Hansen's energetic public demonstrations to push his country into adopting sustainability caused him to get arrested on several occasions.[26]

Inspired by Hansen's research results, Dr. David Suzuki, host of the CBC science program, *The Nature of Things*, also warned his audiences about the dangers of climate change. This encouraged Canada's federal government to initiate a historic meeting.

The First Major International Climate Change Conference

In 1988, Brian Mulroney (Canada's Prime Minister from 1984 to 1993) and Thomas McMillan (Minister of the Environment, 1985–1988) organized a novel international conference, the first of its kind: "The Changing Atmosphere–Implications for Global Security." It was held in Toronto, June 27–30.[27]

The conference's 300 international delegates from 40 countries represented science specialists and high-level political policy makers. A consensus was reached on June 30 that climate change was a reality with dangerous consequences: "The world is entering an era of unprecedented climate change" was the way the conference put it.[28] Bouchard immediately concluded, "We need to cut back on fossil fuels, now."[29]

23 Suzuki (2023a, minute 3.33)
24 James et al. (2022)
25 Wikipedia (n.d)
26 Wikipedia (n.d.) Toronto conference
27 Fenech (2013, Jul 2)
28 (Toronto Conference (n.d.) Wikipedia)
29 Bouchard (2023, minute 0.30)

Looking back on the conference twenty-five years later, a bitterly disappointed Canadian attendee wrote: "Humanity is conducting an unintended, uncontrolled, globally pervasive experiment whose ultimate consequences could be second only to a global nuclear war."[30]

This historic conference made the general public aware of the global concerns over what scientists were predicting about climate change. It also encouraged the growth of alternative industries that would develop "clean" electrical energy sources (e.g., solar, wind, hydrogen, batteries, etc.).

Within Canada today, the province of Alberta leads in wind and solar energy development; yet it is home to Canada's fossil fuel industry; a.k.a. "Big Oil."[31] In the summer of 2023, Alberta's new Premier, Danielle Smith of the United Conservative Party, imposed a seven-month moratorium on new wind and solar projects. The shadow of Big Oil may have passed over Alberta's legislature that day.

This decision will likely cause "billions of dollars in green investments to flow away from Alberta to the United States and to Europe."[32] Such a loss of investment translates into Alberta's loss of jobs for its labor force. However, this was of no concern to Albertan rural voters who "have pushed back against wind and solar projects, arguing they are ugly and take up land."[33] Meanwhile their land today has become severely parched from worsening droughts due to 2023 being the hottest year on record. "Old-timers say they haven't seen soil so dry," Stueck[34] reported. Without appropriate provincial government action, their rural livelihood will soon disappear, which is indeed uglier than wind turbines and solar farms.

Globally, the "top four countries" producing clean energy today are:[35] (a) *solar*: China, Germany, U.S., and India; (b) *wind*: Germany, China, and the UK; (c) *hydrogen*: Australia, U.S., Spain, and Canada; and (d) *battery* production: China accounts for 77% of global sales.

Back in 1989, Bouchard had warned, "If we don't move now [towards clean energy], it will be a disaster. The survival of our species is at stake. I don't want to scare people, but it is a matter of great intervention."[36] But attempts

30 Fenech (2013, Jul 2, website quote)
31 James et al. (2022)
32 Jones & Graney (2023, website quote)
33 Tait & Graney (2023, website quote)
34 Stueck (2024, Jan 10)
35 Google searches (2023)
36 Bouchard (2023b, minute 23.37)

at introducing the public to sustainability (i.e., moving from a profit society to a sustainable one) became ephemeral or problematic due to the mainstream culture in place at those times.

David Suzuki characterized the 1980s with an ideology[37] held by manufacturers, corporations, politicians, and the general public: "There are no limits to what could be done"[38] in terms of: (a) taking resources from the Earth—There's a lot more where that came from; and (b) sending polluting garbage back into the land and oceans—Don't worry, be happy. Simply put, people generally did not like major changes.

Toronto's 1989 conference's news releases stated: "We [the conference delegates] call for a 20 percent reduction in greenhouse gas emissions within 15 years."

Suzuki warned Canada, "It's too late for discussion, and now it's time for action. ...Action must occur at the highest levels of government."[39]

In 1989, Canada's acting minister of the environment, Lucien Bouchard, counseled: "If you want to set up solutions, you need to change your way of life;"[40] by which he meant: switch to sustainable living and begin by immediately proposing a schedule for seriously diminishing the production of fossil fuels.

A Canadian Story

In response, Canada's fossil fuel industry launched a climate-change-denial campaign to avoid cutting back on their exploring, refining, and selling oil and gas. Scientists like David Suzuki were publicly attacked as being a radical environmentalist. Big Oil lobbied provincial and federal governments to support a *greater* production of fossil fuels; for instance, by finishing the mega projects already under development. These included: the west coast liquid natural gas terminal, the Lloydminster refinery, and Newfoundland's Hibernia off-shore drilling platform.[41] In their propaganda, Big Oil focused on the negativity of a subsequent massive worker layoffs that would cause repercussions in Canada's economy. Such a prediction is a famous political half-truth, a typical ploy to oversimplify a complex event (discussed below).

37 In general, ideology refers to a doctrine of how one treats people
38 Suzuki (2023a, minute 10.36)
39 Suzuki (2023b, minute 22.30)
40 Bouchard (2023, Jul 18, minute 22.50)
41 CBC Ideas (2023, minute 24.25)

Perhaps it was the propaganda, but in any case, (acting) Environment Minister Lucien Bouchard changed teams. "I don't think we can stop everything with a radical move. . . .We are also concerned with other aspects of life, like jobs. We have to live with the results of the past. . . .Keep Hibernia and look into the future."[42] The meaning of "look into the future" was not made clear. But rhetorically it *sounded* progressive.

At the same time, Stephen Lewis, Canada's ambassador to the UN for the environment and economics, was the one who had originally suggested, "Reduce CO_2 emissions to 20 percent by 2005." He added:[43] "The policies we have at the moment encourage us to do just the opposite."

In the aftermath of the Toronto conference, a race:[44]

> to transform the world's energy system to make it cleaner and greener is well under way, but while Canada is making some progress, there's a long way to go. . . . the World Economic Forum ranks Canada nineteenth out of 120 countries on its Energy Transition Index.

The forum's report pointed out that Canada needed to diversity its mix of energy sources, strengthen its institutional and regulatory frameworks to create change and be open to innovative technologies. This is indeed a challenge with the jurisdiction squabbles and interfering setbacks initiated by oil rich provinces.

Today, Big Oil representatives lobby governments to continue giving subsidies for fossil-fuel projects. Similarly in 2022, Big Oil pressured the UN's 27th annual "Conference of the Parties" meeting in Egypt not to place oil and gas on the COP's "phase-out" list along with coal.[45][46] Such lobbying is a case of the blind leading the blind in terms of avoiding climate change's tipping point associated with a disastrous ending for humanity as we know it.

Since 1988, there has been an enormous economic price for Canada to pay for these outsiders' corporate resistance to changing the normal way Big Oil has reaped large profits. Three examples illustrate the price paid in 2023:

1. Drought-causing heat waves have ravaged farmers' crops and animal feed, especially in the prairie provinces.[47] Herds have been culled.

42 Bouchard (2023, Jul 18, minutes 27.46–31.37)
43 S. Lewis (2023, minute 17.20)
44 Israelson (2023, Oct 30)
45 Mackell (2022)
46 Details are below in subsection "Lifting the Curtain . . ."
47 Levy (2023)

Families and crop insurance companies have been hit hard.[48] On top of that, farmers' crop insurance premiums have sky-rocketed.
2. Record-breaking wildfires have razed villages and neighborhoods across Canada.
3. Intense thunderstorms punished Nova Scotia with record amounts of rain that became raging currents across a wide swath of Nova Scotia, where torrents washed away roads, bridges, and two children from their family's car, drowning both children.[49]
4. These types of destruction have dramatically increased the cost of property insurance.[50] In fact, for vulnerable locations, insurance companies have terminated people's policies altogether, and these people have trouble securing a new policy from any insurer, or at a price they can afford.[51]
5. Resale value plummeting due to potential risks of increased climate change destruction. It has already affected a quarter of the United States' residential properties. It is a bubble that amounts to the "high hundreds of billions, if not trillions of dollars across the U. S."[52]

For a moment, imagine 1988–1989 in Canada. If the reasons for significantly reducing fossil fuels had prevailed over privileged political influence back then, those Nova Scotia children and other people would still be alive today.

More specific information about the financial debts that climate change has already caused are summarized here:

1. *Seriously cut into a country's economy*. For instance:
   a. "Canada is, on average, experiencing warming at twice the rate of the rest of the world, with Northern Canada heating up at almost three times the global average, according to a new government report."[53]
   b. Eurostats' research established Europe's average cost *per year* (transposed into Canadian dollars) to be CAN$ 21 billion.[54]

48 Harder (2023)
49 CTV News (2023, Jul 23)
50 Saba (2023)
51 Rabson (2023)
52 Vanderclippe (2023, Sep 23)
53 CBC News (2019, Apr 4)
54 Eurostat (2022)

c. Canada, therefore, most likely will significantly exceed this CAN$ 21 billion figure, given that Canada has a higher warming rate.
d. For 2024–2025, the Canadian Climate Institute predicted that climate-induced damages will "slow Canada's economic growth by CAN$ 25 billion annually."[55]

2. *Ruined economies* (detailed below in subsection Climate Change Damage During 2022):

   a. Pakistan's flooding in 2022 killed more than 1,700 people, displaced tens of millions, and caused upward to CAN$ 30 billion in damage.[56]
   b. Since 2021, southern Europe's extraordinary heatwaves, droughts, wildfires and flooding have devastated economies, from Portugal to Greece.[57]
   c. A myriad of documented cases are itemized in the *New York Times* bestseller by David Wallace-Wells, "*The Uninhabitable Earth: Life after Warming*." Spoiler alert! Its content has terrified some readers.

When Bouchard suggested that a shift to clean energy sources will ruin the employment lives of workers in the petroleum industries, he was obviously assuming business as usual: that government leaders would arbitrarily plan and then force a reduced use of fossil fuels, without collaborating with workers' unions in the early planning stages of the transition from a pure-profit society to a reasonable-profit sustainable society.[58]

Suzuki offered a realistic perspective on the adjustments in peoples' lives. "We cannot have all our present conveniences tied to fossil fuel, directly or indirectly."[59] *Collaboration* is the key strategy for *minimizing* both the inconveniences and the temporary financial downturns that are familiar to most of the general public (e.g., the 2008 global recession). While both are inevitable, the alternative is unthinkable.

Suzuki advised: "The message is clear, if humanity is going to survive, we must change the market place where profits rule. Human intelligence can foresee danger and take action to avoid it. If not, humans become the endangered

55 Saunders (2023)
56 Afzal et al. (2022, website quote)
57 World Meteorological Org (2023)
58 Carney (2020b, 2021)
59 Suzuki (2023b, minute 38.10)

species, overtaken by the planet's heat."[60] Internationally renowned economist, Dr. Mark Carney concurs.[61]

Bickering and law suits between, on the one hand, oil producing provinces and on the other hand, the federal government, have been less than productive. The federal government makes broad societal decisions, given the post-normal scientists' evidence and advice. The oil producing provinces have tended to make specific decisions irrespective of the scientific evidence, but in line with their party's ideologies and loyalties. Both types of governments operate the same democracy, but with different duties.

The rising politically motivated tensions between big-oil provinces and Canada's federal government prompted an "enough-is-enough" editorial from one of Canada's national newspapers. The editorial was entitled, "Provinces have climate powers—and duties."[62] For instance, the Alberta government would seem to have acted blindly to *democratic privilege* when they initiated a hiatus in the production of clean energy technologies. With powers comes responsibility to exercise those powers for the greatest good. In a different setting, parents' conversations with their teenage offspring call it "maturity."

The answer to "What action should we take?" is a complex question, because from both sides, action requires reasoned whole-truths and not politically expedient half-truths—policies based on simplistic, arbitrary, self-serving viewpoints. To paraphrase Saunders'[63] way of explaining the complexity in which humans found themselves in 1989, and even more so in 2023:

Many people have come to realize that:

1. The considerable personal and public *costs of* stopping and reversing climate change, *are much less than*
2. What people will pay and lose personally in the long run, if we continue with present policies.

This full truth contradicts the halftruths in the pro-Big-Oil propaganda; for example, Minister Bouchard's position stated above concerning workers' survival with clean energy transitions.

For a number of voters, however, this full truth often gets translated into: Short-term pain for voters means loss of the election by the candidate.

60 Suzuki (2023b, minute 46.40)
61 Carney (2020b, 2021)
62 Editorial Board (2023, Oct. 19)
63 Saunders (2023)

For a political party, that is a dilemma! The full truth is all about what is best for the country.

Since the topic arose in 1988, the dilemma has caused little significant, country-wide, political action on solving climate change. Perhaps this will change as the general public becomes personally moved by the steady increases in human and financial catastrophes, due to the climate-change crisis. This has already become evident to some extent because of the news media coverage of annual meetings that address climate change.

The United Nations Annual Conferences on Climate Change

The United Nations began holding a yearly "Conference of the Parties" (COP) in 1994 in Berlin (COP_{21}).. Its agenda was: "to stabilize the greenhouse gas emissions and to protect the Earth from the threat of climate change."[64][65]

U.S. Vice President Al Gore's 2007 book, *An Inconvenient Truth*,[66] re-popularized the climate change topic. He received serious push back, however, especially from the banking and petroleum industries; an intense instance of the negativity toward change, very apparent within post-normal science concerning climate change.

Publications about major climate change news often refer to COP_{21} (Berlin, 1994), COP_{21} (Paris, 2015), and to COP_{27} (Sharm el-Sheikh, Egypt, 2022). A quick summary of the purposes for each of above three meetings could read like this: (1) COP_{21}: to protect the Earth, (2) COP_{21} to avoid an impending crisis, and (3) COP_{27}: to end the crisis we are in. If there were a climate change report card, COP would receive a failing grade, I submit. My evidence comprises the four following subsections concerning Paris.

Prior to the 2015 Paris COP, anyone who had the mental ability to extrapolate scientifically seemed to have been hijacked by their own personal I-am-the-centre-of-the-universe agenda; very similar to how the sun-centred universe challenged the Renaissance public's I-am-the-centre-of-the-universe ideology as well.

64 Krishnamurthy (2022, website quote)

65 Down to Earth (2022)

66 Gore (2007)

The 2015 Paris Accord (COP_{21})

The Paris meeting brought 196 nations together in an attempt to determine what is required to avert an impending climate change crisis.

The agreements reached at COP_{21} became known as "the Paris accord." It was a major improvement over previous COP meeting outcomes because its anticipated achievements were quantified. The Paris accord reduced today's average global temperature to 2.0°C above pre-Industrial Revolution levels; or more preferably 1.5°C. The Industrial Revolution occurred approximately between 1760 and 1840.

The highly complex calculated results that arrive at those figures are summarized in a graph by Climate Scoreboard.[67] Their graphs show internationally determined contributions of greenhouse gas emissions. The complexity is due to the many assumptions that different scientists try to agree upon. Some came up with a 2.0°C target as being reasonable to avoid reaching a point of no return—a *serious* tipping point of climate change—with temperatures soaring higher and higher each year with no possible way of reversing the world's increasing average temperature.

Climate-Change Tipping Points

Complex mechanisms explaining the "no possible way,"[68] classify the phenomena from weak to serious.[69] An example of a serious tipping point involves the most potent greenhouse gas, methane. Large amounts of it are trapped in the Earth's subterranean permafrost (a large portion of it is located in Canada).[70] Permafrost is frozen moist soil that traps methane produced by biochemical reactions in the soil. During the summer, parts of the permafrost melts. As it does, some trapped methane escapes into the atmosphere, and raises the Earth's global temperature a small amount. During the winter, more permafrost freezes, and the cycle continues.[71]

However, if *all* the permafrost should happen to melt, then all the trapped methane would escape into the atmosphere. This spike of methane would raise the Earth's temperature even higher. If this happens, there is nothing to cool

67 Environmental Century Files (2015)
68 OECD (2023)
69 Carbonbrief (2020)
70 O'Connor et al. (2010)
71 Heginbottom, Dubreuil, & Harker (1995)

the Earth. Then the Earth's temperature steadily rises on its own without stopping; eventually destroying all life on Earth.[72] This information is omitted in Big Oil's half-truths rationalizing the enormous profits it is protecting.

Understandably, some cautious scientists came up with the safer target of 1.5°C, even though it is a more difficult pathway for a country to take.

This was the first of five main outcomes to the COP_{21} meeting in Paris. The other four were: (1) a first universal climate agreement, (2) helping to finance damages experienced by poorer nations, (3) publishing greenhouse countries' gas reduction targets, and (4) agreeing to become carbon neutral (i.e., net-zero balancing between greenhouse gasses entering and exiting the atmosphere) by 2050.

The meeting's hallway atmosphere was tense like a war zone. People were beginning to realize that there may likely be a shorter time to act. Action must be accelerated at a warlike pace. World War Zero is likely eminent.

One participant, Mark Carney, wrote about integrating three types of humanistic "technologies:" Engineering, Politics and Economics; that is, merging the three through holistic reasoning. Carney's 2021 book, *VALUE(S): Building a Better World for All*, shares this conclusion with Kepler's book *Harmony of the Worlds* (i.e., referring to the natural, the humankind, and the humans' spiritual worlds).[73] Carney and Kepler represent two different sets of metaphors in two different eras; but both with a similar purpose—human survival on planet Earth.

In 2022, a major international research questionnaire on sustainability reached over 4,300 university students enrolled in science classes related in some way to environmental issues.[74] The students represent the future decision makers in their own countries. The researchers' purpose was to give these future decision makers "important guidance on how the SDGs ('Sustainable Development Goals') should be promoted in their country."[75] The students were asked to rate 17 United Nations' SDGs in terms of their importance.

Although the ratings changed depending on the country, most students rejected the idea that environmental problems are simply the price of progress. The study's conclusion was optimistic about a second Scientific Revolution occurring to address climate change. Solar electricity was students' popular clean energy source.

72 Carbonbrief (2020)
73 Carney (2021, p. 298)
74 Kleespies & Dierkes (2022)
75 Kleespies & Dierkes, p. 1

The university students' judgments about the most important SDGs appeared to fit into three groups: the social, environmental, and economic components of sustainable development.

The study's conclusion emphasized the need for *strong interconnections* among these three groups; simply put, *holistic thinking*.[76]

In other words, it ***is*** feasible for humankind to achieve harmony between itself and Planet Earth. The smart money today is being invested in sustainable, "green," commercial projects, processes, and products.[77]

Enter COP_{27}, held in Sharm el-Sheikh, Egypt (November 6–18, 2022). It was the UN's 27th attempt to agree upon how to tame climate change. It brought together world leaders in politics, economics, and industry. Their job was to respond to the brutal destruction of lives and property that occurred in 2022 alone; summarized here.

Climate-Change Damage During 2022

Between COP_{17} (Paris) and COP_{27} (Egypt), the world's major carbon polluters did not appear to reduce climate change damages. Au contraire! Four examples of dramatic *increases* in environmental damages caused by climate change are illustrated here.

First, "catastrophic floods of 2022 submerged a third of Pakistan under water."[78] The increased heat from climate change caused glaciers to melt much faster than usual, thereby making rivers to overflow. The increased heat also caused huge hyper-torrential monsoons to form. Both swelling rivers and monsoons made "lakes out of villages and cities, killed more than 1,700 people, displaced tens of millions, and caused upward of $30 billion in damage, with vast swaths of crop-land destroyed and rampant disease,"[79] virtually destroying Pakistan's economy.

Secondly, southern Europe experienced an extraordinary wildfire season; incredibly worse than the preceding 15-year average.[80] There was a large jump in the extent of wildfires in 2022 compared to its average number of wildfires over the previous 15 years.[81] 2022 was an exceptionally hot year.[8283] Threating

76 Hendry (2014, p. 45); Kleespies & Dierkes (2022, p. 7)
77 Carney (2020b); Israelson (2023)
78 Afzal et al. (2022, website quote)
79 Afzal et al. (2022, website quote)
80 ***European*** **F**orest ***Fire*** **I**nformation **S**ystem is EFFIS
81 Davies & Cursino (2022)
82 Routley (2022)
83 Handcock (2022)

fires on the slopes into the major port city of Genoa, Italy, almost jumped the fire breaks into the city.

In 2023, nature's fiery assault continued. On the island of Maui, Hawai´i, several thousand residents and tourists ran for their lives through streets and alleys, as gale-force winds whipped wildfires swiftly across part of the island.[84] It quickly destroyed parts of the centuries-old ocean-front town of Lahaina, trapping at least 100 people to their fiery deaths. Some escaped by running to the ocean, leaping in to swim for their lives. In Canada's Northwest Territories, its capital city of 20,000 was evacuated just before a massive wildfire raged in quickly. No one could believe how fast that fire could move. Climate change is showing us things we've never seen before.

Thirdly, high-temperature records throughout Europe were shattered. As a direct result of the ensuing wildfires, once peoples' 2022 fire insurance claims were processed, many of them discovered they could not afford their increased fire insurance premiums. Who should reimburse them so they can have fire insurance protection?

Fourthly, water became so scarce in areas of Europe that major rivers evaporated into creeks or became totally parched. Some parched riverbeds had donkey carcasses strewn about; some partially consumed by predators. Many farmers had nothing to sell in the town markets.

As a major result, science-technology solutions became *closely* intertwined with economic issues and political policies. This means that a one-dimensional solution (i.e., solely a science-technology solution) that uses reductionist thinking in a largely profit society is simply not feasible. It is a losing strategy.[85]

Solving the planet's climate-change crisis requires changing nature's atmosphere. Thus, nature predominantly dictates the rules of engagement: *holistic multi-dimensional* thinking in a largely *sustainable* context. Science's reductionist thinking will serve as a problem solver if and when such reasoning is required.

As mentioned above, economists and corporations' Chief Executive Officers (CEOs) keep a ledger of costs and profits. But their costs only relate to their business' immediate costs, not to the costs to others: those who have suffered floods, droughts, wildfires, etc. that can be indirectly but rationally traceable to the profit-making actions of those businesses and corporations. Thus, a ledger only tells a half-truth, not the whole-truth of what is happening.[86]

84 CBC News (2023)
85 Gale (2023, May)
86 Cuthand (2023, Aug 26)

"Every rich country today became wealthy, thanks to the exploitation of their fossil fuels; directly from the ground or by selling its products. The elite countries' idea of green energy for the world's poor, is incapable of transforming lives."[87] Some compensation for poor countries by the rich countries occurs through a UN international fund called "loss and damage" (L&D).

The ethics are straight forward. Consider what happens in a city neighbourhood at a community-league baseball game, when a home-run fly ball happens to break a nearby house window. The two teams or the league's organization *automatically* take care of the damage. No dodging this common-sense ethic.

Internationally, countries responsible for gaseous emissions or rain-forest clear cutting that do damage to their distant neighbours' health and livelihood have the same common-sense responsibilities. Two differences are:

1. The window scenario is on a simple 1:1 proportion—the teams or baseball organization pay for a definable cost of the glass and its installation. Moreover, people are watching.
2. Responsibilities for making reparations for the effects of either greenhouse gas emissions or clear-cutting need to be proportional to: (a) the country's relative emissions or clear cut logging (such data are already available); and (b) to the estimate of the damage caused to others.[88] Therefore, one should not be surprised that COPs seem to end in a standoff on these two issues; because few people are watching a meeting so far away.

"A geopolitical storm is still powering a global energy crisis."[89] COP_{27} is a prime example of trying to address directly the entrenched ideology of a profit- society.

Highlights from COP_{27}, Sharm el-Sheikh, Egypt

In 2022, the United Nations COP_{27} meeting in Sharm el-Sheikh, Egypt (November 3–18) brought world politicians, economists, and industry together to address the current crisis of climate change and the responsibilities of the world's rich countries.

Political leaders from poor countries publicly called out the Big Oil delegation at the conference, and the leaders of rich countries: "demanding that Big

87 Lomborg (2022)
88 Ayed (2022)
89 Board (2022)

Oil pay up for damages inflicted on the poor economies."[90] The payments are a direct measurable result of the carbon emissions sent into the atmosphere by each rich country. Promises made during Paris' COP_{21} had not been adequately kept. Yet the industries' speeches at COP_{27} sounded as if there were no problems collecting on promises to help.

Big Oil's compensation record is no better than the record of the Renaissance Emperors, Kings and Dukes who were notorious for not paying people like Kepler for what was promised during the first scientific revolution. Parallels between Renaissance Emperors and Big Oil today are both striking and disappointing.[91]

"Leaders from poor countries criticized wealthy governments and oil companies for driving global warming . . .demanding that they pay up for damages being inflicted on their economies."[92]"Small island states call on oil firms to shell out some of their profits to compensate them for losing their shorelines and beaches to rising sea levels."[93] These treasures use to support their economy, reliant on tourism.

At COP meetings, "Mention their Loss and Damage obligations, and most of the delegates from large polluting countries—the ones whose carbon emissions are largely responsible for baking the planet in 2022—they change the subject or leave the room."[94]

As if that is not bad enough, there is a country that is: (a) the world's largest emitter of greenhouse gases, (b) accounts for about 27% of global emission, but (c) has no official obligations to the Loss and Damage fund. This country privately negotiated a "developing country" status with the World Trade Organization.[95] The country is China.[96]

A business group named "Pathways Alliance" is made up of oil companies such as: Suncor Energy Corp, Imperial Oil Ltd, Canadian Natural Resources, etc., in Alberta, Canada. "They have collectively pledged to work together to reduce the sector's absolute emissions by 22 million tons [of carbon] by 2030."[97] This includes the Athabasca Oil Sands (a.k.a. "the tar sands"). Mr. Dilling,

90 James et al. (2022)
91 McGreal (2021)
92 James et al. (2022)
93 James et al. (2022)
94 Board (2022)
95 Board (2022)
96 Reguly (2022a)
97 Stephenson (2022)

President of Pathways Alliance, "admitted that the sector now realizes its very future depends on achieving the pledge. We have hit that TSN Turning Point;"[98] a term familiar to sport fans. (The expression was first used by The Sports Network to describe a video highlight of a play that by itself, influenced the outcome of the game the most.)

Another Dilling pledge is an "additional $7.6 billion investment in other major emissions reduction projects and technologies. Pathways has not yet begun their carbon capture project."[99] A growing chorus of voices has been calling for the industry to move faster with its climate pledges—especially given the record profits that energy companies had earned in 2022. Thus, they certainly do have the money to pay their "fix-a-broken-window" debt.

"Dilling acknowledged the criticism but said the decision to go ahead can't be made until regulatory approval for the project is granted (by the government), a milestone that's at least two years out."[100] No mention is made of having dragged their feet applying for such approval.

However, both: (a) "TSN Turning Points" in investment pledges, and (b) unsubstantiated excuses for non-action; are no game-changer to solving the climate change crisis. It would appear that only worldwide intense public pressure is required to make the Dilling team act ethically. In this context, I intend that "ethically" translates into: moving from a profit society to a sustainable society with reasonable profits.

Dilling's Pathways Alliance has a "history of failed promises when it comes to supporting the environment and seriously addressing climate change."[101] Inspiring such public pressure is Mark Carney's agenda (Chapter 2).[102] One of the largest members of Pathways Alliance, Suncor Energy Inc., has a new CEO, Rich Kruger, who identified his new goal: "to focus on the business of low-cost production and profits rather than the long-term energy transition."[103] He plans to steer Suncor Energy away from "future focused, clean and low-carbon energy pursuits." Kruger sounds like the spokesperson for both the profit society and climate change deniers. In the war of humanity against itself, he is speaking for the destruction of humanity.

98 Stephenson (2022)
99 Lomborg (2022, website quote)
100 Stephenson (2022)
101 Stephenson (2022); Lomborg (2022); James et al. (2022)
102 Carney (2021)
103 Graney (2023b, Aug 16)

The Paris COP_{21} in 2015 warned the world about the devastation and expense of not lowering the Earth's average temperature to the 1.5°C above pre-industrial levels, before 2050. In 2015, countries generally agreed to avoid that impending crisis.

Glasgow's (UK) COP_{26} in 2021 essentially repeated a similar message. The Sharm el-Sheikh (Egypt) COP_{27} meeting was a paraphrase of Glasgow, more or less. But it subtly indicated that we were indeed experiencing the crisis that the contributing industries promised to avoid all along.[104] "Already by the fall of 2022, the temperatures have increased by 1.1°C since 2015, and we are projected to blow past 1.5°C, without implementing swift and deep cuts to emissions within this decade (i.e., by 2030).[105]

Bright spots in the COP_{27}'s summit's final text are both familiar and express a "supernova" aura: they are staged for bright media headlines, then they dim soon after.

Although a motion to strengthen the loss and damage fund barely passed at the COP_{27} meeting, a subcommittee will have to work out its details before the fund will pay out any money. These details will be debated at COP_{28}. And the meetings will continue to negotiate over who promises to pay a certain amount.

Canada's Environment and Natural Resources (ENR) has a "Climate Change" department. The following quote comes from their report on their contribution to COP_{27}.[106]

> The Government of Canada established the 2030 Emissions Reduction Plan on March 29, 2022 to provide a credible roadmap to achieve Canada's 2030 target and put us on a path towards net-zero emissions by 2050. Canada is also finalizing our first National Adaptation Strategy in 2022 and delivering on our five-year (2021–2026) $5.3 billion international climate finance commitment.

One cannot avoid the conclusion that this accurate description appears to mimic the genre of press releases from Big Oil. For instance, there is a plan for the future, but nothing about the degree to which it met its plan that preceded this one. Chapter 5 presents data that indicate E.N.R.'s present plan is already outdated.

Anticipating that Canada's 2023 federal budget will prioritize solving the climate change crisis, the Green Budget Coalition's research concluded that

104 Daigle et al. (2022)
105 Daigle et al. (2022)
106 E.N.R. (n.d.)

$20.9 billion over five years is required to "address: the climate and biodiversity crises, create sustainable jobs and ensure equitable prosperity and well-being for everyone in Canada."[107] This includes: Zero-Emissions Electricity, Renovation Wave, Land and Ocean Protection, Sustainable Agriculture, and Environmental Justice; in other words, the "what should happen" model.

Lifting the Curtain on the Highlights of Egypt's COP_{27}

To be a savvy world citizen about climate change, you should be able to identify the numerous "Wizards of Oz" stories occupying the COP ("Conference of the Parties") and beyond. The Globe and Mail's European Bureau Chief, Eric Reguly, has lifted the curtain on this collection of Wizards.[108] Here is what you need to know.

The UN's COP has four funds now.

1. *Mitigation Fund*: To make greenhouse gas emissions less severe or damaging; to prevent their escape into the atmosphere; and to lessen the conversion of rain forests into agriculture fields; etc.
2. *Adaptation*: To invent new processes or technologies that will remove greenhouse gases from the atmosphere.
3. *Coal Phaseout*: Stop burning coal; the result of COP_{26}.
4. *Loss and Damage* (discussed above): Reimburse poor countries that have suffered disastrous environmental damages due to climate change traced to the economies of rich industrial countries. It is supposed to paid in U.S. dollars.

The above four funds capture the COP in theory. You should certainly be aware of what *really* happens. It's not pretty.[109]

A fairly popular self-promotion gimmick by some countries must be exposed. They donate double their committed amount to one fund with much media coverage, but quietly lessen their contribution by the same amount to another fund.

At Egypt's COP_{27}, COP teams from 200 countries participated for the 2-week meeting, stretched longer in order to finalize an agreement, and avoid the embarrassment of not doing so. Public pressure was the motivator.

107 Green Budget Coalition (n.d.)
108 Reguly (2022b)
109 Reguly (2022b)

There were 600 oil and gas lobbyists paid by many oil companies, "Big Oil." They persuaded COP_{27} to let them continue new explorations and further development of their current business models, even though the Earth's climate change had entered a crisis status. They were successful in defeating a motion to add "oil and gas" to the Coal Phaseout fund. Despite unusually frank discussions over "phasing out oil, gas and coal, the end result was yet another COP without formal recognition that Big Oil is driving the climate crisis."[110] When will Big Oil's charade end? Seasoned journalists described COP_{27} as "*an exercise in organized hypocrisy.*"[111]

Big Oil's success at hypocrisy will certainly make it more difficult to reach the 1.5°C goal agreed upon at the Paris COP_{21} meeting. Thanks to this difficulty, potential payouts from richer countries to the UN's Loss and Damage fund will need to cover the predictable greater damage costs. These increased costs ultimately become greater taxes for citizens to pay.

These tax payers will not be amused. Meanwhile, big profits are made by Big Oil. Schemes of "corporate welfare" by Big Oil are revealed, directly linking some corporate profits to tax payers.

It is time to point out that the word "commitment" has two very different meanings in the halls of a COP meeting. First, there is the bonified obligation to carry out what was promised. The second meaning: "Maybe we'll pay but maybe not."

The United Nations' COP's record over the years shows clearly that the task of dealing with the climate change crisis is beyond the capability of this international political-industrial complex. Part of the comprehensive picture is the fact that democratic countries rely on voters, and investment brokers rely on shareholders who invest in the petroleum companies in the first place. These players need to exert pressure for sustainability, Mark Carney's agenda. Unfortunately, positive intentions for action submerge in the "cultural ocean" of a profit society,[112] represented by a few but influential prevaricating procrastinators.

110 Reguly (2022b)
111 Mackell (2022, website quote)
112 Carney (2020a)

· 4 ·

A SUSTAINABILITY REVOLUTION: WHAT TO AVOID

Humanity has gone through a Scientific Revolution (about 1543–1686 CE), an Industrial Revolution (about 1760–1840), and is currently dealing with the consequences of the Digital Revolution (1960s plus). Unexpectedly, humanity has recently been threatened with eventual extinction by going beyond a serious tipping points of climate change.

On the other hand, a take away from Chapter 3 would be: we face a climate-change crisis. Chapter 4 describes how and why this was allowed to happen. By learning that, we can plan to avoid it in the future.

History has reached the point where another revolution is overdue. In the context of economics, Mark Carney pointed out in Chapter 2 that a Sustainability Revolution could still create sustainable global societies out of today's societies that are run by the ideology of making maximum profits.

Such a Sustainability Revolution needs to be *broader in scope* than just: economics, UN style of politics, post-normal science and technology, *reductionist* ways of reasoning, and the Eurocentric traditions for making sense out of the universe; most of which were inherited from the Scientific Revolution.

Chapter 4 critiques these "Eurocentric" conventions in order to reveal what needs to be left behind before a Sustainable Revolution can move forward.

A Sustainable Revolution must embrace worldwide innovations appropriate to cleansing the atmosphere. This international collaboration needs to be organized on a global scale among: post-normal natural scientists and engineers, social scientists (particularly economists),[1] mathematicians specializing in artificial intelligence;[2] plus: informed, effective, collaborative, and politically savvy politicians competent at understanding the nature of science and its societal interactions (NOS).

But small-scale projects of such collaboration never did reach their global goal, in spite of some splendid attempts. What happened? Chapter 4 describes how humanity let it happen. Pick from it, what you think should be avoided in the near future, before it is too late.

The Petroleum Industry's Critical Position

First, just a reminder or a quick update: greenhouse gases include: methane (from cattle flatulence and garbage dumps), nitrous oxide (from car and truck exhausts), and carbon dioxide (from burning wood, coal, oil, gasoline, diesel, jet fuel, propane, natural gas, etc.)—all substances together causing the largest proportion of atmospheric contamination that raises the average global temperature. and most containing carbon atoms; hence, the measurement term "tons of carbon."

The petroleum industry contributes a large amount of money in taxes to the economy of oil-producing countries. And for all countries, comes sales tax on that petroleum sold.

Business people will tell you that sometimes it is shrewd to take a temporary loss in profits in order to make greater gains *down the road*. Big Oil does not seem to behave that way when it comes to climate change.[3] Big Oil's preoccupation with keeping investors happy in the short term is a motivation to always show a profit. It becomes: a habit; a part of one's personality and self-identity; or even an addiction. All are difficult to change, unless a wildfire, a flood, or a heatwave attacks: your family, your home, your income, your pets, or you personally.

Big Oil's strategy to sidestep the climate change crisis is to word their messages in a way that keeps the public impressed and investors satisfied with

1 Carney (2021)
2 Borovik (2017)
3 Lomborg (2022); Jones & Bradshaw (2022)

half-truths.[4] What *lies down the road* (i.e., the full truth) is basically ignored; for example: (1) the savings gained by installing lower costing electrical sources of energy, should be subtracted from the initial cost; (2) personal savings paid out to costly bills or taxes to pay for reconstructing infrastructures damaged by the increasing destructive effects of climate change. Catherine McKenna (Former Canadian federal minister of environment and climate change) calls their half truths "cheap talk." For instance:[5]

> Regrettably, Suncor, Shell and BP announced plans to slow-walk clean energy and pump more fossil fuels – the main cause of the climate crisis. They chose to spend the vast majority of their record profits on shareholder dividends and executive compensation while *investing a small fraction in the clean energy transition*. Incredibly, at the same time, oil sands companies demand that Canadian taxpayers spend *even more* to subsidize their carbon capture projects.

The 2015 Paris COP_{21}'s main achievement was the agreement to control the atmosphere's average temperature so it gets no higher than 1.5°C above pre- industrial levels.

At the 2021 Glasgow COP_{26} meeting, Mark Carney, UN Special Envoy for Climate Change, organized the most powerful group of economists into an alliance. Their goal was to reach "net-zero" by 2050 or sooner. As a reminder, "Net-zero" means to absorb greenhouse gases *from* the atmosphere at the same rate as they *enter* the atmosphere due to human activity.

Both processes work toward a net-zero balance. But achieving such a balance is not the end of the journey. More greenhouse gases need to be removed in order to reach the 1.5°C mark.

"A ***carbon sink*** is anything that absorbs more carbon from the atmosphere than it releases; for example, *plants*, the *ocean* and *soil*. In contrast, a carbon source is anything that releases more carbon into the atmosphere than it absorbs; for example, the burning of fossil fuels or volcanic eruptions."[6]

Eating red meat should be added to ClientEarth's list of carbon sources due to the methane released: (a) in cattle's and lamb's flatulence, (b) from garbage dumps, and (c) from permafrost (see subsection "The 2015 Paris Accord in Ch. 3). Methane causes much more atmospheric warming per kilo than carbon

4 Willis (2022)
5 McKenna (2023)
6 ClientEarth Communications (2020, website quote)

dioxide does."[7] However, methane amounts to only 11% of the greenhouse gases in our atmosphere, compared to carbon dioxide's 79%.

Reaching Net-Zero Naturally and By Engineering

Nature provides its own carbon storage mechanisms. In the domain of *plants*, forests contribute most to drawing carbon dioxide *from* the atmosphere. "Amazonia hosts the Earth's largest tropical forests and has been shown to be an important carbon sink during recent decades. This carbon sink seems to be in decline, however, as a result of deforestation and climate change."[8] Amazonia is discussed just below in Chapter 5, section "Harmony to Achieve Net-Zero."

Wetlands are the most efficient sink, because they store "more carbon than any other ecosystem,"[9] on a per acre (hectare) basis. In addition, wetlands help prevent floods and droughts. Other ecosystems such as coast lines and forests are natural sinks. Wetlands are reduced by urban sprawl and by agriculture development. Conservation of prairielands and wetlands are good sustainable practices toward net-zero.

Many industries and university departments are doing net-zero research and development (R.&D.) to reach the UN's target date of 2050. This requires a great deal of R.&D.

For example, similar to most countries worldwide, Canadian university R.&D. projects include: tracking climate change impacts (University of Alberta); unlocking the potential of plant-based materials (University of B.C.); bringing human and artificial intelligence together (University of Montreal,—artificial intelligence, for example, can already identify worst hit neighborhoods caused by hurricanes etc.[10]); accelerating Canada's shift to a low-carbon economy (University of Victoria); and mapping climate futures (Wilfred Laurier University).[11] These examples illustrate a Sustainability Revolution's multidimensional and holistic features.

These R.&D. projects need funding from governments (i.e., tax payers), loans from banks, and public investments (e.g., in the stock market).

Public investments generated their own separate industry long ago. Specialist economists ("financial planners") advise clients on how to manage their

7 E.P.A. (n.d.)
8 Gatti, Basso, Miller, et al. (2021, website quote)
9 Government of Canada (2023)
10 Stokel-Walker (2022)
11 Randall Anthony Communications (2022)

investment assets. Investors hope their assets will increase in value over time.[12] That is how the system works.

The UN's G.F.A.N.Z. group (Glasgow Financial Alliance for Net-Zero) initiated and led by Mark Carney, is made up of special interest groups in economics, such as:[13]

- Net-Zero Asset Managers initiative,
- Net-Zero Financial Services Providers Alliance,
- Net-Zero Banking Alliance, and
- Net-Zero Insurance Alliance.

The G.F.A.N.Z. has the power to influence financial planners, and special interest groups, to invest only in industries or businesses that have initiated sustainable practices (i.e., "green" assets).

Therefore, perhaps G.F.A.N.Z. is a potential game-changer to help establish a Sustainability Revolution. By increasing the G.F.A.N.Z. collaboration into a critically large group, along with other major players in society, net-zero criteria and the 1.5°C will have a chance to be met in time to avoid a runaway climate-change disaster. If so, the sanity side of an ethical sustainable world will have won the war.

Threats to sustainable world order are currently showing themselves in the Net-Zero Banking Alliance group, however.[14] Banks have profited well by loaning large amounts of money to Big Oil for their exploration, drilling, and pipeline activities. The large reserves of oil maintained at the present time, may likely be largely out of date by 2050, and therefore only a small amount of oil is needed. There go the Banks' large profits associated with Big Oil's loans for their exploring for, and drilling new wells.

Banks especially worry over the threat of legal action against them for dropping their investments in high carbon-emitting companies (mostly the "oil and gas, and mining"[15] industries). Such legal agreements sat well with climate change deniers.

But obviously the banks made bad business decisions back then. And now they want others to pay for *their* bad decisions by paying for the damages caused by the climate change that was the consequence of their bad decisions in the

12 Ganti (2022)
13 Carney (n.d.)
14 Jones & Bradshaw (2022)
15 Jones & Bradshaw (2022)

first place. Simply put, they want to play neighborhood baseball, but won't pay for the windows they break. Of course, the economic sector of society is much more complicated than the baseball metaphor. Those complications, however, are often turned into reasons to procrastinate, that is a convenient "profit-society fortress" for the banking industry.

In summary, the goal of a Sustainability Revolution will be a transition from a profit society to a sustainable society that offers the opportunity for rational profits, without the threat of climate-change tipping points.

Getting the Data Oneself

Perhaps you are curious about how well your own country is meeting its targets for the year 2030, on its way to achieving the new 2035 net-zero and 1.5°C expectations.[16] If so, there are agencies in most countries that collect the appropriate impartial data and report it on the internet. For instance, the C.C.I. (Canadian Climate Institute) tracks the implementation of targets, plans, and policies in Canada.[17]

They identify both climate policy successes and priorities for improvement. Two recent notes from C.C.I. are:

1. "Hitting Canada's 2030 climate targets hinges on the rapid and effective implementation of a new policy that cuts emissions across the entire oil and gas sector."
2. "Top Canadian companies aren't yet accounting for all the pollution released from the products they sell, potentially exposing them to both economic and reputational risks."[18]

This type of neutral information differs in its intent and trustworthiness from a CEO's presentation to potential investors. For example, "CEO says it's critical to balance fossil-fuel projects with renewable energy objectives."[19] A thoughtful savvy citizen needs to interpret this quote *in the context* of its source. For instance, this statement is the usual profit-society lingo (perhaps the company was selling off some of its assets). The context is certainly not a time-sensitive,

16 E.C.C.C . (**Environment and Climate Change Canada** (2024)
17 C.C.L. (2022)
18 C.C.L. (2022)
19 Willis (2022)

sustainable society, climate-change-crisis type of context—the context the world is in at that moment.

As reported from the COP_{27}, Big Oil's pledges and predictions are simply mirages to entice the public to invest in their assets; for example, TC Energy Corp. It also acts as subliminal messaging to encourage the public to resist the transition into a sustainable society.

At the same time, the United Nations is encouraging international economists to show leadership in reorganizing[20] the politicians, the industrialists, the post-normal scientists, engineers, mathematicians, etc. They must address Chapter 2's section, "The World of Economics," in which an agenda remodels the current open-market's ultimate control over society into a more equitable sustainable society.

At the Toronto Metropolitan University, engineer Venkatesh commented, "Utilities are looking towards us to create that transition in an *affordable manner.*"[21] Big Oil seems to think an *affordable transition* is one in which Big Oil continues to receive the large *profits* that it feels it is entitled to, while a seismic *shift* to a sustainable society is trying to take place. But Big Oil is playing the half-truth game. The full-truth includes the fact that the world economies are already beginning to require companies to certify that the product was produced with a net-zero electricity grid.[22]

It is simply not possible to have *both* (i.e., acquiring large profits and participating in the process of becoming sustainable) during the *short* time before reaching a serious climate-change tipping point.

Yet again, this reality offers Big Oil a thin excuse to procrastinate while the worlds' damage-and-debt meter rises exponentially due to the worsening effects of climate change.

Supported by the Competitive Enterprise Institute, Big Oil and other related industries, back in the 1990s, decided to fight the evidence that predicted disastrous outcomes from climate change. It turns out that those decisions made by certain industries were bad business decisions. The ball is now in their court. Why should the public pay (indirectly through taxation) for bad business decisions by Big Oil and others?

Is there a chance the current dissenting bankers will go ahead and support G.F.A.N.Z.? Or will they continue to hold the world economies hostage, by threatening financial harm if G.R.A.N.Z. does not back away from its agenda

20 Carney (2021)
21 Venkatesh (2022, p. CFI4)
22 Wilkinson (2023, May 29)

of a speedy-as-possible transition to a sustainable society. "What harm?" you might ask. By acting in a way that bankrupts the global financial system into a recession—the 1930s all over again but from a different reason.

The UN's COP movement has demonstrated that humanity is at war with itself; rich countries against poor countries. The northern hemisphere against the southern hemisphere. Areas suffering horrendous yearly droughts create huge human migrations into neighboring cities and countries, causing all kinds of stress.

Ideological wars continue unsettled between those who are pro-*petroleum* sources of energy and those supporting *sustainable (clean)* electrical sources of energy (e.g., wind, solar, hydrogen, thermal etc.). In the heat of this "war," an accusation of "witchcraft" (Renaissance European style) will arise against the sustainable side who promote fixing climate change's force as soon as can be, for the sake of today's children and grandchildren.

The sustainability side needs to recruit reinforcements to match the social media and financial force of the climate-change deniers or the tipping-point deniers. These reinforcements hold promise for all countries to achieve net-zero before it is too late. Because sustainability-minded citizens live among us, there is hope.

For instance, the sustainable-minded Thomas McMillan served as Canada's Minister of the Environment from 1985 to 1988 in Prime Minister Brian Mulroney's government.

From Action to Inaction and to Action Again

As described in Chapter 3, Environment Minister Thomas McMillan and Prime Minister Brian Mulroney in 1988 had worked hard for Canada to lead the world in bringing countries together at Toronto to rationally examine the challenges of climate change that lay ahead.

Because McMillan became busy teaching for the first semester of 1989 at Yale university of 1989, Canada's acting (temporary) Minister of the Environment, Lucien Bouchard, immediately asserted that the first step to avoid a planetary disaster was to abandon quickly a large portion of fossil fuel production.

However, it took over 21 years (until 2009) for successive federal governments in Canada to "commit to ending inefficient fossil fuel subsidies that 'encourage wasteful consumption, distort markets, impede investment in clean energy sources, and undermine efforts to deal with climate change.'"[23]

23 Gunton, Wheeler, & Tienhaara (2023, website quote)

Moreover, other economic cuts must also be carried out in a rational manner. Finally in 2022, the federal government pledged to adopt the 2035 deadline for achieving a net-zero electricity grid. This means that oil production will need to make plans to decrease quickly.

Electricity, as a source of energy, will replace a large portion of fossil fuels (i.e., coal, natural gas, and gasoline). When this replacement is sufficient to ensure that a net-zero greenhouse gas emission has been attained, the situation is called "*a net-zero electricity grid*," a new popular expression in today's discussions about climate change and international commerce.

Over the years since the 1988 (the year of the Toronto international conference on the existence of climate change), Big Oil has made a series of *business decisions* when influencing government policies and the UN's COP meetings *against* phasing out petroleum products to a bare minimum. According to the International Institute for Sustainable Development (I.I.S.D.), those bad decisions are the reason that today Canada's oil industry is not prepared for an impending drop in the global demand for petroleum products, "as the world scrambles to reduce greenhouse gases emissions to avoid the worst effects of climate change."[24]

Like any Canadian business, executive decision makers must take responsibility for the mistakes they make, and not shift the financial consequences onto the public's income taxes to help cover Big Oil's losses. These costs are for the industry's investors to share. Shareholders will lose some money. They will not be amused. Likely some highly paid CEO's heads will roll in the board rooms. This is inevitable; a *normal* casualty in the world of business: short-term pain for a long-term gain.

According to the UN Intergovernmental Panel on Climate Change (I.P.C.C.),[25] it is almost too late to avoid climate change's serious tipping points of no return. The year 2050 was an interim target date for cooling the planet's average temperature to 1.5°C above what it was just before the Industrial Revolution. As you will read in Chapter 5's Introduction, the I.P.C.C.'s latest climate information indicates that the target date of 2050 should be lowered to about 2035. Major political decisions must be made immediately and are currently under discussion between Canada's federal and provincial governments.

24 Feinstein (2023, Jun 28); Graney (2023a, Jun 27)
25 I.P.C.C. (2023, Mar 20)

The Action Surfaces

Fortunately in 2022, the Canada Energy Regulator (C.E.R.) switched from being a Big Oil supporter to recognizing the severe and costly problem of the climate change crisis. C.E.R. recently published their research by mapping out in technical details "what a net-zero world would look like in how Canadians produce and use energy."[26] To achieve net-zero greenhouse gas emission by 2050, the CER produced three different complex scenarios: Canada, Global, and Current Measures. Their endpoint detailed descriptions equally applicable to 2035.[27]

A net-zero electrical grid draws on: wind, solar, hydrogen, geothermal, bioenergy, nuclear, and hydro energy sources, as well as on a three-step process called "carbon capture, utilization and storage" (C.C.U.S.).[28] "Fossil fuels will still play an important part in Canada's energy system, particularly in industry facilities outfitted with C.C.U.S. . . .The electricity sector is projected to achieve net-zero emissions by 2035."[29] Big Oil cried "foul" in its half-truth sort of way, without sharing the whole truth on what is "foul."

The Government of Canada (G.C.) had shown timely leadership in creating Toronto's 1988 international conference on climate change. Today, the G.C. is finally showing the political fortitude for an *improved second effort at leadership, by establishing a collaborative pathway to a sustainable energy future.* But the G.C. has only a short time-line to make that effort succeed.

The key decisions on technical matters have been completed by C.E.R. Very recently, the G.C. has made historic and mindful decisions in the best interest of Canadian citizens' future, and specifically of Canada's workforce. Both must cope with the *world's* transition from a mainly petroleum-based energy source, to mainly an electricity-based energy source.

An unbreakable commitment by the G.C. has been made to achieve the 2035 net-zero target:[30] "for the Canadian economy to achieve 40–45 percent emissions reductions"[31] and:[32]

26 C.E.R. (2023, website quote)
27 C.E.R. (2023)
28 C.E.R. (2023, website quote)
29 C.E.R. (2023, website quote)
30 Wilkinson (May, 2023)
31 Gov't of Canada (2022, Mar, website quote)
32 Gov't of Canada (2022, Mar)

> to cut more pollution, to create more good jobs, and to support a healthier economy and environment. The plan sets out a vision, goals and near-term targets across five interconnected systems already in place: (1) *disaster resilience*, (2) *health and wellbeing*, (3) *nature and biodiversity*, (4) *infrastructure*, *and* (5) *economy and workers*. Projects will be developed and financed within these areas.

$5.3 billion during 2021–2026 was made available; $315 million of which was "allocated for projects from civil society."[33] Severe push back arose from the oil producing provinces.

However, Canadian firms looking to export products such as critical minerals, hydrogen, potash, steel and aluminum are increasingly being held to account for their products' carbon content,[34] defined by their greenhouse emissions used during their production of those products (i.e., full-truth accounting).

The G.C. could establish a new source of money to help pay for emissions reduction grants that it awards.[35] This new tax could be partnered with the present sales tax, but based on the *carbon content* of a manufactured object for export. Columnist Doug Cuthand called it a "carbon tax." It would encourage people to purchase low-carbon items. Perhaps the name is rather abstract. "Climate-change crisis tax" could make people feel better. To accomplish a public "I told you so" reprimand, call it the "Blame-the-science-deniers tax."

Following 1989, Canadian governments procrastinated by putting the petroleum industry's short-term benefits ahead of Canadian citizens' long-term needs; as well as the world economy, which was rapidly changing. "This green transition is occurring principally in the U.S., China and Europe."[36]

Canada's Minister of Natural Resources, Jonathan Wilkinson, recently pointed out:[37]

> Decarbonization of the electricity grid is an environmental imperative. ...Being able to provide clean power to industry is increasingly the 'the price of entry' into the marketing of products internationally. . . .Saskatchewan's BHP, for example, is building the $7.5 billion Jansen potash mine. BHP's recent announcement regarding this project underscored the importance of emission reduction to its business.
>
> . . .It's not just Canada that is working to achieve a net-zero grid by 2035. It is in fact every G7 country. Just last week, the U.S. released its regulatory approach

33 Gov't of Canada (2023, Jan, website quote)
34 Wilkinson (2023, May)
35 Cuthand (2023, Aug)
36 Mason (2023, website quote)
37 Wilkinson (2023, website quote)

> to achieving a net-zero grid by 2035. Our American friends are doing this for both environmental and economic reasons.

The U.S. is several years ahead of Canada on this issue. International business contracts with Saskatchewan manufacturers and producers are switching to the U.S. because of its modern climate-change policies.[38]

This U.S. success originated with legislation passed by President Biden in 2022. It "offered billions of dollars in subsidies for green energy projects. . . . Clean energy is moving faster than many people realize."[39] According to the executive director of the International Energy Agency (I.E.A.), of the US$ 2.8 trillion energy budget this year, it "invested about 61 percent (US$ 1.7 trillion) of the U.S. energy budget in: wind, solar, electric vehicles and batteries."[40]

"Of course, the U.S. still relies heavily on oil and gas. So does China. But they are both moving aggressively to get to a place where fossil fuels are a thing of the past. According to reports, China is set to double its capacity of green energy by 2025, reaching its 2030 goal five years ahead of schedule."[41]

And the Action Continues

"Six Canadian provinces are on track to meet the 2035 goal because more than 90 percent of their hydro and nuclear power generation is already emissions-free."[42]

Three provinces produce 93% of Canada's oil: Alberta (headquarters for Big Oil), Saskatchewan, and Newfoundland and Labrador. The first two provinces account for most of it. Their provincial governments vehemently oppose the federal government's new climate-change position that favors moving quickly to sustainable energy sources, defended by federal ministers Steven Guilbeault and Jonathan Wilkinson.[43]

The Saskatchewan government's pushback against the federal government's climate-change plans and targets, for instance, delivered by science- denying political leaders who treat the climate change crisis as nothing but "whimsical ideology talk." Saskatchewan's Premier Scott Moe, assured his place on the Dunning-Kruger Effect's board of directors when he defended his government's

38 Wilkinson (2023)
39 Wilkinson (2023, website quote)
40 Mason (2023, website quote)
41 Mason (2023, website quote)
42 Neis (2023, Aug 16, p. A1)
43 Gov't of Canada (n.d.)

rationale for not switching quickly to renewable energy sources: "We will not risk plunging our homes, our schools, our hospitals . . . into the cold and darkness because of the *ideological whims of others*."[44]

Premier Moe's rhetoric was upstaged by his Minister of Justice, "who once wrote in a newspaper column that scientific proof of global warming was '*witchcraft thinking*'."[45]

South-west Saskatchewan farmers, ruined by the sequence of droughts during 2022 and 2023,[46] did not seem to connect the dots between their devastating droughts and their Saskatchewan government's resistance to acknowledging the scientific evidence-based cause of those droughts. These same rural citizens have normally voted the same "Saskatchewan Party" back into office.

Saskatchewan newspaper columnist, Phil Tank,[47] challenged Scott Moe's claim that the Federal government's action plans will cost Saskatchewan $111 billion. "So much focus is grabbed by the upfront costs, it eclipses the benefits or the money saved." In other words, Moe was only telling a half-truth, rather than the full truth.

Tank continued: "It costs money to increase energy efficiency in a home, but you benefit financially in the long term."[48] For instance, a Saskatchewan family invested $6,500 to replace their gas water heater by installing an electric water heater system. Electricity is much cheaper than gas in Saskatchewan. Upon instillation, the family's monthly gas bill suddenly decreased. Over about 10 years, these monthly savings added up to CAN $6,500. And the low monthly bills continued to save the family money compared to what they would have otherwise been paying. Living more sustainably has an upfront cost, but many replacement-sustainable purchases will pay for themselves over time.[49] This important point was buried by people who are trying to manipulate the public into thinking with only a half-truth, rather than with the whole picture.

Strong pushback aimed at the federal government's climate-change plans also came from Alberta, the oil capital of Canada. Strangely, it is also the leading province in the development of wind and solar electrical energy sources. As mentioned previously, Danielle Smith became Premier of Alberta in the fall of 2022. Apparently, she wanted to alter the climate-change topic from what is best for the

44 Cited by Mandryk (2023, May 19, p. A5)
45 Cited in Mandryk (2023, May 19, p. A5)
46 Harder (2023)
47 Tank (2022)
48 Tank (2022)
49 Mason (2023); Wilkinson (2023)

citizens of Alberta, to which government (provincial or federal) can legally decide what path to take. In other words, it became simply a political power issue!

It came as a shock in the summer of 2023, when Smith suddenly placed a seven-month moratorium on all wind and solar energy projects; most likely an appeal to rural voters who usually see wind turbines and solar panels as "butt-ugly and a scam."[50]

Smith's surprise move rattled the targeted companies.[51] They had invested CAN$ 5 billion over the past four years in Alberta. These companies will now be thinking about moving their CAN$ 3.7 billion future investments and 4,560 jobs to the US or Europe.[52] Similarly: "the directors of 15 projects currently in Alberta's approvals queue, and the 91 projects at an early development stage,"[53] may be having second thoughts about doing business with Alberta. The most positive remark about Smith's risky decision was that it was an amateur move,[54] which did not seem to serve Alberta's economy and work force well.

It will take more time, diplomacy, and bold decisions to achieve a resolution to how best to solve our climate-change crisis.

As promised in Chapter 1, this book's next subsection will identify some key players and their ideas that correspond to the ideas that Kepler and his natural philosopher colleagues had when they *changed the culture* of Renaissance Europe. As a result, they created the beginning of modern science.

Modern Scientists, Engineers, Economists, Politicians and Corporate Leaders Act

Today these people are educated to develop or interpret, facts and explanations about controlling the physical, economic and societal worlds we live in. Their assumptions include:

1. Follow the data and decide what ideas the data are telling you. Nature does not lie. It speaks truth through its actions. Economists' ideology changed in the eighteenth century to the "market never lies"—a.k.a. laissez-faire[55][56] (Chapter 3).

50 Mason (2023, website quote)
51 Jones & Graney (2023)
52 Simakov (2023)
53 Nickel (2023, website quote)
54 Simakov (2023)
55 Carney (2021)
56 Wikipedia

2. Separate those ideas by organizing them into logical groups or categories. To do that, your mind uses *reductionist thinking*—it reduces ideas into general categories.
3. Look for *relationships* among those categories. That should help you make more sense out of the data, thereby creating more abstract categories to help you think about one category at a time.
4. Politicians are key players in the action, too. But they tend to begin, not with data, but with an ideology (in the sense of a doctrine about how people should be treated). Canadian politicians tend to emphasize the goal: *peace, order, and good government*. Practicality forces them to focus on winning the next election, however. Politicians who demonstrate sophistication, grace and competence tend to make thoughtful decisions; that is, making decisions based on the best ideas and the most thorough data.
5. Citizens who are: international entrepreneurs, local business people, or investors; also tend to hold strongly to the ideology of a pure profit society (Chapter 2). It has been a strong part of our culture since the eighteenth century. It has obviously been the unexpected source of the climate change crisis of today.

To understand this mixture of people involved today in ideological warfare is a challenge for the general public trying to make sense out of what to do about the climate change crisis. Chapter 4 ends by giving special attention to Canada's quagmire of conflicting political and economic interests today—ideological warfare.

Corporate Action

One well-known successful business leader in the petroleum industry is Mr. François Poirier, CEO of TC Energy Corp, based in Calgary. The corporation was established in 1951 to lay natural gas pipelines from the prairies to eastern Canada. In December 2022, Poirier met with people who had purchased shares in TC Energy Corp, and with a reporter at the meeting.[57] These shareholders were concerned about the escalating costs of a new gas pipeline project being built in British Columbia.

57 Willis (2022)

Andrew Willis' news item on the meeting and interview offered insights into Big Oil's strategies. For example, Poirier supported the establishment's profit society stance while acknowledging aspects of a sustainable society, but without mentioning the time-sensitive status of global warming's current crisis and imminent total disaster if it reaches a serious tipping point. When Poirier reminded his audience of "the goals of TC Energy Corp's biggest client—the utilities that power the economy,"[58] his message reflected the Madison Avenue advertising style promoting TC Energy Corp. itself, by associating it with the powerful entities of North America's economy.

Poirier understands that his CEO job "involves balancing the priorities of current fossil fuel-focused projects and a 30-year plan to become a low-carbon company."[59] This statement assumes or overlooks important details: (a) how "low" with respect to 1.5°C, and to "net-zero"? and (b) in the meantime, who pays for the costs of those "30 years" (i.e., until 2053) of mega debts from property destruction, infrastructure reconstruction, and agricultural failures (from storms and droughts); plus an increasing number of torturous deaths from wildfires and agonizing drownings from horrendous storms?

CEO's debit-credit style ledgers are too simplistic to capture these costs of dealing with a climate-change crisis for 30 years. Simplistic becomes a strategy.

Poirier emphasized TC Energy Corp's "resiliency" during those 30 years.[60] Try telling that to the victims of today's climate change crisis. His resiliency talk belongs to 1988 when there was an honest chance of attaining sufficient sustainability by 2018. But Poirier's people fought it. The destructive years 2022 and 2023 were Sunday picnics compared to what an aggrieved Earth has in store for humanity. That is a scientific *fact.*

On the other hand, Poirier told the reporter, Andrew Willis, "It's critical to balance fossil-fuel projects with renewable energy objectives."[61] That is not a fact; but instead, a linguistic play on the words "critical" and "balance," expressing a Big Oil *ideology.*

Poirier continued, "Looking down the road, the company's E.S.G-friendly growth is expected to resonate with investors."[62] ("ESG" is Big Oil's popular feel-good acronym for "Environmental, Social, and Governance.") The acronym also means "socially responsible investing" and "sustainable investing."[63]

58 Willis (2022)
59 Willis (2022)
60 Willis (2022)
61 Willis (2022)
62 Willis (2022)
63 12:eleven, 2023

However, that ESG-friendly-growth train left the station in 1988, which makes catching up more challenging.

Since 1988, there has been a consistent impasse among four groups of citizens: government, industry, science and the general public. Scientific evidence and conclusions have been lost on those people who generally embrace a pure-profit culture, but have been supported by people who generally want to live in a non-climate change culture.

These groups are interrelated: (a) *industry* must follow the laws of the land; which are: (b) passed by *federal* and *provincial governments*; on (c) the advice of *scientists, engineers* and *economists*; and (d) governments are elected by the *general public* who decides as a group who wins an election. With those dynamics in mind, an editorial by a national newspaper (*The Globe and Mail*) advised: "Oil companies need a push on climate."[64]

That puts the general public in charge: (a) by how they vote when electing the politicians; and (b) as consumers, the choices they make regarding the type of energy—green electricity-based or petroleum-based energy—it took to manufacture the purchased item. That is where governments may be able to make the biggest difference. They can legislate that each product's use of energy during its manufacturing will be marked on the package. That allows consumers to "vote" personally, for or against climate change's destructive powers, on the basis of the item's type of energy and how much is used during the item's manufacturing. The higher the energy count, the greater the chance of destruction.

Using this type of friendly pressure, governments are now able to make their move to confront climate change and still get re-elected. For the sake of the children, it is time to change cultures and promote a more sustainable society with reasonable profits, before humanity blindly slips over a serious tipping point, pushed by the fossil fuel industry.

That "time to change" urgently arrived on the vernal equinox of 2023. It caused the UN Secretary-General, António Guterres, to announce in utter frustration that enough is enough. And then in public, he hit the climate-change panic button. "UN chief urges faster shift to 'net zero'."

According to "a top UN Intergovernmental Panel on Climate Change (I.P.C.C.)," we are dangerously on a course to exceed a tipping point to increasingly higher temperatures. António Guterres expressed the warning this

64 Globe & Mail (2022)

way: "Humanity is on thin ice—and that ice is melting fast. Our world needs climate action on all fronts—everything, everywhere all at once."

He called an end to new fossil fuel exploration and for rich countries to quit coal, oil and gas by 2040. "Nearly two-thirds of carbon pollution must be slashed by 2035" at no higher a temperature of about 1.5°C above pre-industrial average temperatures, rather than by 2050. He "urged developed countries to commit to 'net zero' by the earlier date of around 2040."

According to the I.P.C.C., emissions must be halved by the mid-2030s if the world is to have any chance of limiting temperature rise to 1.5°C above pre-industrial levels, a key target enshrined in the Paris accord. Beyond doubt, the climate-change crisis was upon us.

In Canada, upon learning the legitimate UN panic-button episode, rumors immediately circulated around Environment and Climate Change Canada (E.C.C.C.) that an annual $3 billion will be available to support a Municipal Climate Action Fund; a case of "Think globally, act locally."

· 5 ·

GETTING IT RIGHT

Introduction

It is time to replace the ideological war with some paradigm-shifting sustainable action that draws on peoples' sense of harmony. This, and along with marginalizing those prevaricating procrastinators who, on the vernal equinox of 2023, caused the UN Secretary-General, António Guterres, to announce in utter frustration that enough is enough[1] (Chapter 2). And then in public, he hit the panic button. "UN chief urges faster shift to 'net zero'."[2]

Therefore, the first topic in Chapter 5 concerns some foundational understandings upon which Indigenous languages are built. This also serves as background for Chapter 6's call for the collaboration between Indigenous and Western (Eurocentric) nations to confront the climate-change crisis we now face.

1 Semeniuk & Stueck (2023)
2 CA Finance Yahoo (2023, Mar 20, website quote)

What Cultural Ideas Do Indigenous Cultures Tend to Share

This chapter highlights some features of Indigenous cultures you need to be acquainted with in order to understand what is about to take place where you live to confront the climate-change crisis. These ideas describe (but do not prescribe) key features of Indigenous cultures. Language is the first topic.

Languages are windows into a peoples' worldview. As you recall, Indigenous worlds are in constant motion (flux). That feature is bound to be reflected in the language their culture developed: a language of change and motion. Consequently, Indigenous languages are action-oriented; that is, verb-based.[3] The world that Indigenous Elders inhabit is in "constant flux."[4] It can change unpredictably and it usually does. You can either read about it yourself, cited in the footnotes below; or you can listen to Indigenous Elders or Knowledge Holders talk about their language. Leroy Little Bear has several informative presentations on YouTube.[5]

In contrast, Eurocentric languages tend to be noun-based;[6] the language reflects a different worldview.

Therefore, when a non-Indigenous author describes Indigenous thoughts using a European language, the author should respectfully use European expressions: (a) that mean what the original *Indigenous* expression conveyed specifically; plus (b) convey a viewpoint consistent with the Indigenous person's worldview. If that English expression sounds odd, so be it.

In this situation, the non-Indigenous author is being given the opportunity to experience what some Indigenous people experience when listening to English. Thus, the noun-based expression "Indigenous science" does not represent an Indigenous person's deep meaning (e.g., the world is always on the move). How can we turn "Indigenous science" into a verb-based expression that obviously reflects an Indigenous deep meaning?

In a 2011 book, *Bridging Cultures: Indigenous and Scientific Ways of Knowing Nature*, Herman Michell (Indigenous) and I (a Eurocentric person) translated "science" (a noun) into "ways of living in nature" (IWLN)[7] (a verb-based descriptor).

3 Little Bear (2000, p. 78); pc.
4 Little Bear (2000, p. 78); (pc.)
5 Little Bear (2011, May 9).
6 Cajete (2000, pp. 27, 72); pc.)
7 Aikenhead & Michell (2011, p. 65)

Terminology is very important. The English words "coming-to-know"[8] or "coming-to-knowing," sound odd, but they convey an Indigenous meaning: to learn *in-depth* while forging a strong relationship with what is being learned. It also entails using more than one's brain. The heart, the body and/or the soul usually become involved to some extent.

In a project where Indigenous and non-Indigenous participants collaborate, the more they come-to-know, the more effective the project's outcome will be.

As previously mentioned, the word "science" has connotations that many Indigenous people identify as negative. For example, the Western habit of thinking in a reductionist way with many dichotomous statements sprinkled throughout.

Negativity also arises when the Eurocentric meaning of "science" forces some Indigenous people to think in a reductionist way, *at the expense of* marginalizing their Indigenous holistic way of thinking. Any Indigenous person thinking holistically would be considered "living in a good way [i.e., a harmonious way] with nature,"[9] or "in nature,"[10] according to their community.

About 20 years ago, I learned that forcing an Indigenous speaking person to use the word "science" in the context of "Indigenous science," can be a subtle form of colonization. Prominent Mi'Kmaw scholar, Dr. Marie Battiste, called the act "cognitive imperialism."[11]

Hendry captured this negative experience from an Indigenous perspective when she wrote: "the English language is a straight-jacket which forces [Indigenous] minds into a world of objects, categories and restrictive logic."[12] I concur. At the same time, Indigenous languages have the potential to be more sophisticated and vivid than English, French, German, etc.

For example, Elder Leroy Little Bear, retired professor from Harvard University's Native Studies Program, contends that quantum mechanics in theoretical physics is more accurately explained in Indigenous languages than in English. Perhaps it is the Indigenous assumption that their universe is in constant flux, which is also assumed when working with quantum mechanics.

The expression, "Ways of living in Mother Earth," is a synonym for "IWLIN." "Ways of living" conveys an action. The plural, "ways," connotes

8 Cajete (2000, p. 110)

9 Aikenhead & Michell (2011)

10 E.g., the Pearson Canada textbook series "*Saskatchewan Science*" (n.d.)

11 Battiste (1986, p. 23); (pc.)

12 Hendry (2014, p. 127)

that differences exist among groups of Indigenous peoples. Up until now in this book, I have been using "IWLN" to embrace both knowledge and wisdom. Now, I shall be explicit when I focus on wisdom alone.

> Leroy Little Bear, a member of the Blood Tribe of the Blackfoot Confederacy (Canada), speaks about science through the lens of Native American culture and teachings. He outlines several tenets of this lens and discusses how we might use them to overcome obstacles reached in the scientific community. These tenets are constant flux, renewal, and place. [13]

He has given us many YouTube lectures. Find them on a Google search; for example:

(a) "What Can First Nations People Say About Quantum Mechanics?" Elder Little Bear.[14]
(b) BanffTalk's: "Indigenous Knowledge and Western Science" Deals with topics such as: "What Is Science?" "Foundational Basis of Native Thought." "Energy Waves." and "Everything is Animate." Elder Little Bear.[15]
(c) Cajete, G. (n.d.): Indigenous knowledge and Western Science (youtube)[16] xxx

Elder Leroy, the philosopher and teacher with a lovely sense of humour, which I discovered the first time I innocently asked him, "What time is it, Leroy?"

He shook his head slowly, while a hint of a cheeky smile warmed his face, and he replied, "You don't really want to know what time it is; you want to know what time it isn't." From there we toured science's rectilinear time and Indigenous cyclical time, and which ones are found in nature. Oh, I said to myself, if we can construct our concept of time, we really do have a hand in constructing our reality to some degree, by participating in evolution!

That will be in a very positive direction if key political decision makers collaborate in policy making with people establishing a Sustainability Revolution.

13 Little Bear (2011)
14 Little Bear (2005)
15 Little Bear (2011)
16 Cajete (n.d.)

Connecting Indigenous IWLN with Policy-Making

To acquire group recognition from a key department within a federal government is an unspoken expectation of a group's acceptance to play a significant role in co-managing the Sustainability Revolution. A parallel occurrence in two adjacent countries will shed light on what a country's Sustainability Revolution may become. For brevity and because both have only begun, each will be summarized.

Canadian First Nations

The coordinated, government-funded project, "Weaving Indigenous Science into Environmental Policy-Making,"[17] is centred at the University of Manitoba, Canada, and administered by Prof. Myrle Ballard of the Anishinaabe First Nation. When in Ottawa, she serves as the Director of "Indigenous Science," officially housed within the federal government's department of "Environment and Climate Change Canada" (E.C.C.C.) in Ottawa. It had a pavilion at COP_{27} in Egypt (Chapter 3).

For Prof. Ballard personally, the term "Indigenous science" meant:[18]

> [A] science of the way of knowing the land. In other words, it is a way of knowing the water, the air, the sky, the horizon; everything about a particular place on Earth. Their knowledge of the weather patterns, their knowledge of how species migrate. It is this knowledge that has enabled Indigenous people to survive.

Myrle Ballard's project does not teach people how to fish, for example. Instead, it is about creating a sustainable society in Canada, *a core ideological characteristic of First Nations' cultures.*

On the other hand, most scientists and government officials, such as those who work in E.C.C.C., are trained at school and university to assume there is only one science—their Western culture's science (i.e., its specific knowledge). Such scientists may not understand the expression "living in a good way with nature" as an Indigenous translation of "wisdom." Consequently, to avoid misunderstandings at the E.C.C.C., Prof. Ballard used "science" for reasons of

17 Sanders (2022)
18 Sanders (website quote)

accurate political communications with her colleagues at the E.C.C.C. Neither does she mention "Eurocentric science," the more accurate expression, because she is politically wise.

Unlike *noun-based* English, the *verb-based* Anishinaabemowin language has embedded in it, scientific-like environment-management information and tools. For instance, streams are given names that tell you: (a) the natural way its water flows, and (b) when the local fish spawn. These verbs indicate the present state of what scientists call "the ecosystem." Parts of some Anishinaabemowin words also indicate "the way it was before, to the present time."[19] Often, one Indigenous word requires a whole English sentence to translate.

To inform the policy makers at the E.C.C.C., she communicates in a three-part strategy: "bridging, braiding and weaving."

1. *Bridging* means raising awareness about "Indigenous science" (i.e., Indigenous ways of living in nature) within the government.
2. *Braiding* is when Western scientists work together on research with Indigenous communities on the land. This would be an excellent culture immersion for the non-Indigenous participants; as Indigenous knowledge and wisdom are braided with Western science (i.e., two-eyed seeing; the next general topic below).
3. *Weaving* happens when the E.C.C.C. department starts to intersperse Indigenous and Western sciences together for better-informed decision-making.

Professor Ballard has been doing research for about 15 years with Indigenous partners in the Canadian Arctic and subarctic. Most recently, they have studied the impact of climate change on polar bears.

"The bear biologists within our team learned tremendously from listening to Elders' stories and narratives," she said.[20] "It's just been beautiful to see how it's just different parts of a puzzle. [Western] science doesn't know it all; Inuit don't know it all, either. And by putting those pieces together, then you just have a more rich and fuller picture of what's going on" in your constructed reality.

Prof. Ballard's discussion is an excellent introduction to the next section in this book, "Two-Eyed Seeing," a topic to explore that follows in this book's next subsection about the United States federal government's newly minted parallel

19 Sanders (2022, website quote)
20 Sanders (2022, website quote)

relationship between the Federal Government offices in Washington D.C. and American Indians.

Two-Eyed Seeing

Two-eyed seeing is an action, like "braiding Indigenous knowledges and Western science"[21] together. A person deals with two fundamentally different cultural languages, ways of living, wisdoms, and worldviews.[22] The idea of two-eyed seeing was created in Nova Scotia, Canada, by Mi´kmaw Elder Albert Marshall. It draws on the strengths of both cultures' ways of understanding, in order to: (a) make sense out of something, (b) solve a problem, or (c) participate in the other culture. Some Indigenous people say, "Walk in both worlds."[23]

When Western (Eurocentric) people interact with an Indigenous culture, they are bound: to find unexpected dualities; to puzzle over paradoxes; or to be amazed at a unity found in diversity.[24] These discoveries are not familiar to Eurocentric-minded people, yet they are generally familiar to Indigenous people.

Maryboy and colleagues explained why.[25] Science "is rooted in the historical development of Cartesian philosophy," which subconsciously seeped into the Europeans' noun-based, linear, dichotomous, and reductionist worldview. Indigenous worldviews, however, also embrace some opposite attributes to the Eurocentric ones stated just above.

Indigenous peoples tend to catch on very quickly to two-eyed seeing because their own culture deals with contradictions and paradoxes fairly smoothly. Similarly, a non-dichotomy can be envisioned to accompany two-eyed seeing: think of one eye connected to the brain only, while the other eye is connected to the brain, heart, body and soul.

The two lenses of *two*-eyed seeing: (a) are *not* dichotomous; and (b) *cannot* be interspersed at random (out of context), because this would either make no sense, or a person would risk being seen as appropriating from Indigenous cultures. The braiding metaphor describes how the two ways of making sense of the world can be *amalgamated* into one entity while each part maintains its individual integrity.[26]

21 C.B.I.K.S (n.d.)
22 Marshall (2007)
23 Aikenhead & Michell (2011, p. 141)
24 Maryboy et al. (2006); (pc.)
25 (Maryboy et al., 2006, p. 4)
26 Dion (2009)

When being braided together, Indigenous ways of living in nature (IWLN) and Eurocentric sciences must maintain their individual integrity (i.e., an individual braid),[27] while acknowledging commonalities and differences between IWLN and "Eurocentric sciences."[28]

Both (a) "sustainable ways of living," and (b) "holistic ways of thinking," affirm an Indigenous holistic harmony by integrating two senses of harmony: its sensual and spiritual dimensions (e.g., governed by music, and governed by what touches one's soul; respectively). By tradition, this harmony between Indigenous peoples and planet Earth is practiced daily. Indigenous Elders teach and support their people to live harmoniously with Mother Earth; in both senses of "harmonious:" sustainable and holistic.[29]

That is the reason Indigenous peoples have survived the worst oppressive physical and emotional conditions on Earth over tens of thousands of years.

During Europe's Age of Exploration (sixteenth and seventeenth centuries) of Turtle Island (North America), most early explorers were not intelligent or wise enough to learn the languages and cultures of the Indigenous civilizations. In many cases, Indigenous people helped those settlers survive the winters.[30] In return, however, the Europeans began to colonize them, by taking their land to build farms and colonies. The Europeans were told by their religious and political leaders: If you find non-Christian land, it is yours for the taking; in the name of your country or by authority of the European *Doctrine of Discovery*.

Today, Indigenous Elders and Knowledge Holders will again help "the settlers;" but this time it is dealing with climate change. The Sustainability Revolution needs to engage people, for example, like Indigenous physicist Phillip Duran,[31] whose book, *The Condor and the Eagle*, successfully searches for a new scientific worldview that unites the heart and mind.

We are fortunate that Elders will openly share their ways of living in nature and their wisdom with others in need. The non-Indigenous are in need!

As you discovered above, Elder Leroy Little Bear, of the Blackfoot Confederation, Alberta,[32] has produced several excellent videos that accomplish such sharing.[33]

27 Garroutte (1999)
28 Aikenhead & Michell (2011, pp. 115–120)
29 Hendry (2014, p. 63)
30 Battiste (2013); (pc.)
31 Duran (2013); (pc.)
32 Little Bear (2000)
33 Little Bear (2011, video); (pc.)

The non-Indigenous leaders who will help organize the Sustainability Revolution could learn much from a culture immersion with any Indigenous Elder or Knowledge Holder, to help prepare for a much-improved version of the Sustainability Revolution.

No religious proselytizing takes place during culture immersions. Instead, you experience the evidence of their culture, and you decide yourself what you want to make your own. Outcomes for Eurocentric leaders to cultural immersions include participants using the following three actions:

1. Drawing on values and wisdom of the Indigenous nations, for a particular situation;
2. Drawing on values and knowledge of Indigenous scientists/engineers, for a particular situation.
3. Expanding one's brain-only interactions with others, to embrace a four-dimensional (brain, heart, body, and soul) interacting.

Judiciously combining these three actions exemplifies two-eyed seeing; drawing on the best of both cultures.[34]

To gain an Elder's assistance, a pathway of reconciliation[35] and collaboration with Indigenous peoples[36] will be the rational choice for establishing a successful Sustainability Revolution.[37] For those who wish to work with an Elder, help is at hand in Herman Michell's book, *Working with Elders and Indigenous Knowledge Systems*.[38] He is from the Woodlands Cree First Nation, Saskatchewan, Canada.

Dr. Greg Cajete, a Tewa American Indian from the Santa Clara Pueblo, New Mexico, has also devoted his life to making Indigenous ways of understanding the world accessible to Turtle Island's (North America) educators. A popular introduction to the topic is his *Native Science: Natural Laws of Interdependence*.[39] His book, *Explorations in Sustainable Living*,[40] makes clear connections between a movement like a Sustainability Revolution and Indigenous wisdom.[41]

34 Hatcher et al. (2009); Marshall (2007); Little Bear (2012, pc.)
35 Michell (2017, pc.)
36 Aikenhead (2020)
37 Cuthand (2023, A4)
38 Michell (2011); (pc.)
39 Cajete (2000)
40 Cajete (1999); (pc.)
41 Cajete (2015); (pc.)

A Diné (Navajo) astronomer, Dr. Nancy Maryboy, introduces her culture through the window of the night sky. Her book, *Sharing the Skies: Navajo Astronomy*, is an example. She audio-recorded a live 1.5-hour group presentation, "Two-Eyed Seeing: Navajo (Diné) Astronomy & N.A.S.A. Moon to Mars,"[42] Also, a similar session was produced for Ojibwe, Dakota/Lakota, Mayan, Hawaiian, and African groups.

The Diné Nation's knowledge and wisdom encompasses contradictions and paradoxes in their way of reasoning.[43]

A number of Indigenous nations on Turtle Island symbolically represent the *interdependence* of: humans, plants, animals, Mother Earth, the Sun, Moon, and night sky by their medicine wheels.

The status among the groups: humans, plants and animals; is usually shared equally. Some Indigenous communities hold humans *less important* than the plants and animals. But all three are less important than Mother Earth.[44]

The four parts of the Medicine Wheel are not separate independent categories like scientific categories usually are.[45] The four parts can interact among themselves. The Medicine Wheels' colors and orientation change from nation to nation. The number *four* has spiritual significance on Turtle Island. To view a Plains Cree Medicine wheel and an Elder's detailed tour of its many meanings, visit Sharon Meyer's website, "Kôhkum's [Grandmother's] Gathering."[46]

All languages use metaphors—expressions that have their own deep meanings, but are connected to other quite different meanings. Some metaphors used by scientists and by Indigenous peoples are very different on the surface, but at a deeper level, some have a similar meaning expressed in different ways.

Consider, for instance, Einstein's formula, $E = mc^2$. It represents the scientific idea that everything in the whole universe is made from just energy (E) and mass (m), an extreme example of reductionist reasoning. The equation expresses the *relationship* between the two. The equation becomes a science metaphor for describing the whole universe. In other words: everything in the universe is related through its energy and mass.

Consider a phrase often spoken on Turtle Island at the end of an Indigenous prayer to Creator. Its English translation is, "All my relations."[47] It represents

42 Begay, Maryboy, et al. (2021)
43 Maryboy et al. (2006); (pc.)
44 Little Bear (pc.)
45 Meyer (n.d.) (pc.). Kôhkum's Gathering
46 Meyer (n.d.)
47 Talaga (2018)

the holistic idea that everybody is related to everything in the universe, so you treat them all as you would treat your relatives. This is an excellent example of how a feature of an Indigenous cultural naturally involves an Indigenous person in sustainability.

Your relatives will include plants and animals, as well as fire, earth, wind, and water (four parts to the Medicine Wheel); without which humans could not survive. But in reverse, plants and animals would do very well if humans disappeared from Mother Earth. Hence, some consequently say that humans have less status in the universe. In most Indigenous cultures, humans are also related to Grandmother Sun, Grandfather Moon, and the stars that serve as resting places for a person who has passed away.

This present section, "Two-Eyed Seeing," has been researched in depth and reported at length in such books and videos as those mentioned above, authored by Indigenous writers. Here are three books by *non*-Indigenous authors. When the authors were welcomed into Indigenous cultures, they made their journey into two-eyed seeing, and have written about it as non-Indigenous authors. Some of the non-Indigenous participants in a Sustainability Revolution will be making similar journeys.

Joy Hendrey is a non-Indigenous, retired anthropologist from Oxford Brookes University, U.K. Her research took place during international travels to visit the Ainu people of Japan, the Māori iwi of Aotearoa New Zealand, the Aboriginal families of Australia, and the First Nations and Métis of Canada. Her book's title is: "*Science and Sustainability: Learning from Indigenous Wisdom.*"[48] It could serve well as a textbook for a university course that focusses on a Sustainability Enlightenment.

Another is Wade Davis' *The Wayfinders: Why Ancient Wisdom Matters in the Modern World.*[49] He concentrates on Eastern Africa, Columbia, Peru, Polynesia, and Australia.

Co-authored by a non-Indigenous and an Indigenous writer, the third book, "*Bridging Cultures: Indigenous and Scientific Ways of Knowing Nature,*"[50] is about Turtle Island.[51]

Not only were these authors mindful to refer to understandings as wisdom, they wrote in terms of Indigenous ontologies and verb-based languages.

48 Hendry (2014)
49 Davis (2009)
50 Aikenhead & Michell (2011)
51 Maryboy (2011)

Harmony to Achieve Net-Zero

The more effective the diverse membership of the Sustainability Revolution becomes at crossing cultural bridges, the faster they will reach net-zero. Their *effectiveness* will help them engage better in two-eyed seeing, within their own field of expertise. This effectiveness increases depending how deeply the non-Indigenous members understand Indigenous cultures. They may pass through three phases, beginning at the surface level: an *ability to identify an Indigenous person*, by their language, dancing regalia, storytelling, engaging in certain activities on the land, and making certain arts.

The *second* phase is typical *folk cultural knowledge* gained by non-Indigenous people from: reading, watching TV, experiencing a short cultural immersion, or having an Indigenous acquaintance or buddy at work. However, beware of public stereotypes.

And lastly, *deep cultural knowledge* includes: wisdom, values, and thought patterns, usually not easily noticed by non-Indigenous people. Such knowledge is learned by participating in some Indigenous activities and ceremonies; for instance, a two-day culture immersion.[52] This leads to establishing relationships with Indigenous people.

Cultural roles among Indigenous relationships can differ from community to community. To what degree are non-Indigenous people aware of holding a deep understanding of Indigenous cultures. The depth varies according to how much it involves: the brain, the heart, the body, and the spirit. When all four are involved, the non-Indigenous person's learning process is often referred to as "coming-to-know."[53]

In both Indigenous or non-Indigenous cultures, sustainable living requires a person to value harmony. For example, Western scientists, economists, politicians etc. will need harmonious two-eyed-seeing relationships with each other, in order to control climate change in a timely way. "Ensuring that resource managers understand and respect multiple ways of knowing will improve stakeholder collaboration engagement across cultures."[54]

Political engagement is essential. Little Pine Cree First Nation journalist, Doug Cuthand, pointed out the example of Brazil. It is home to the tribes of a million Indigenous people who sustainably care for its vast Amazon rain forest.

52 Meyer, Aikenhead, Cardinal, et al. (2019, pp. 104–105)

53 Cajete (2000. p. 110); (pc.)

54 Weiss et al. (2013. p. 285)

It is "the lungs of the planet that captures carbon dioxide from the atmosphere and releases oxygen back into it."[55] The gem of sustainability.

The former President of Brazil, Bolsonaro (2019–2022, officially removed from office), opened the Amazon rainforest to plundering by lawless ranchers, loggers, and miners. It had become Brazil's wild west.

A new president nicknamed "Lula" was elected in 2022[56] and immediately appointed an Indigenous environmentalist (a net-zero activist) to his cabinet as Minister of Indian Affairs. Lula's priority is land management: a positive effect on returning the planet's average temperature not more than 1.5°C of its early nineteenth century level. Politicians who are already moving significantly towards a sustainable society have an essential role to play in establishing the Sustainability Revolution.

As mentioned above in the section "Indigenous Wisdom," Indigenous people had helped the original European settlers on Turtle Island survive the frigid winters.[57] Similarly, today Indigenous Elders and Knowledge Holders have a central role to play in using two-eyed seeing to help[58] non-Indigenous community leaders develop a sustainable perspective.

The objective now is to reach a net-zero equilibrium and then the 1.5°C goal by 2035, thereby curtailing further current-day disastrous effects due to climate change.

A first step should be for non-Indigenous participants in a Sustainability Era to develop a reconciliatory relationship with some Indigenous leaders and participants. Non-Indigenous people must initiate the first step. That is protocol, which entails acting in a reciprocal manner. Find out the appropriate gift to offer the Elder you want to visit, in order to initiate the process of asking for help. Upon first meeting him or her, let them know you have a request, as you extend your gift to them in your left hand, while you shake hands with your right hand.

Elders and Knowledge Holders will share their knowledge and wisdom to move toward a sustainable society to help neutralize climate change as quickly as we can to begin to implement a sustainable era.

Canadian hockey legend Wayne Gretsky used to say that he didn't chase the puck, but rather anticipated where it was going. Likewise, people need to

55 Cuthand (2023), (pc.)
56 Cuthand (2023, p. A4)
57 Battiste (2013); (pc.)
58 Truth and Reconciliation Commission, 2016

stop living in the past (a profit society) and begin the future (a sustainable society) now.

The next two subsections of this chapter illustrate how a R.&D. team can work towards climate change's new 2035 goal on a small scale. Two government sponsored projects are summarized here to illustrate: (1) helpful government policy making; and (2) sustainability, reconciliation, and renewal of Indigenous-Settler relationships. The subsections show what can be accomplished within energetic, forward thinking, and reasonably financed Sustainability minded groups who have been involved with Indigenous sustainable ways of living in Mother Earth. Their plans are not just speculation. Other projects are quietly forming today in several locations in many countries. Build on their successes.

This is two-eyed seeing, of course. If these encounters were carried out on Indigenous land in order to exemplify "living in a good way," it could be called "a culture immersion."

In the fall of 2022, Prof. Ballard involved government scientists in learning from Indigenous scholars and Elders known for their expertise on environmental issues. As a result, the ECCC department (Environment and Climate Change Canada) has been successful at improving Canada's action on climate change.

Choosing appropriate English terminology is very important for establishing a positive relationship with Indigenous people. The greater we "come-to-know"[59] the local Indigenous way of making sense out of their world, (i.e., how an Indigenous person typically makes sense of it), the more effective is the outcome of a project that "bridges," or "weaves" or "*braids*" the two cultures together.[60]

To contribute to the Sustainability Revolution, the university's Commerce, Mathematics, and Political Science faculties need to develop or join multidimensional projects.

Sustainability, Reconciliation and Renewal

At the University of Saskatchewan's School of Environment and Sustainability, Dr. Maureen Reed defines Indigenous wisdom as: "common values, community engagement, transdisciplinary collaboration."[61]

59 This is a translation of an Indigenous idea of learning something in depth. It evokes a personal attachment.

60 Sanders (2022)

61 Reed (2022, p. CFI 3)

In general, her research explores the transition from a profit society to a sustainable society for non-Indigenous communities. Her explorations do not follow an economist's point of view, which is Mark Carney's jurisdiction.[62]

Instead, she works with "the sustainable knowledge of local and Indigenous peoples who live and work in communities and landscapes"[63] in this natural sustainable way.

Dr. Reed and Prof. Ballard have small scale versions of what is needed to reach net-zero. Their results will be scaled up to formulate specific strategies to help manage the climate change crisis. Their multidimensional approach is through citizen action located in communities. This illustrates a few features of a Sustainability Revolution.

These two start-up university projects are less than the "tip-of-the-iceberg" of what is occurring at universities around the world. There are many engineering R.&D. projects underway at this time, looking for sources of electrical energy as alternatives to petroleum.

"To be successful in our transition towards sustainability, we have to address the worries, hopes, concerns and limitations that people see in that transition," Dr. Reed advises.[64]

Conclusion

We need to redefine "humanity's war against itself" mentioned earlier in this chapter. We need to recruit knowledgeable and wise people who understand Mother Earth from a sustainable perspective (e.g., Indigenous Elders, Knowledge Holders, and culture-conscious Indigenous scientists, economists, engineers, politicians, mathematicians, etc.). The best of two worldviews, Eurocentric and Indigenous, need to be braided together through two-eyed seeing to solve more quickly one of humanity's most serious global problems.

Marie Battiste,[65] demonstrated how Indigenous knowledge and wisdom are intricately connected to the immediate place where people live: their land, horizons, water, sky (day and night), and ecosystems. This is called

62 Carney (2021)
63 Reed (2022, p. CFI 3)
64 Reed (2022, p. CFI)
65 Battiste (2013); (pc.)

"place-based" or "land-based" knowledge.[66] Place-based or land-based knowledge valued in Indigenous cultures, contrasts with generalized abstract knowledge valued in Eurocentric cultures. There are many differences between these two positions. When people connect with the land, they feel open to its emotional, physical and spiritual essences.[67]

66 Michell et al. (2008)

67 Little Bear (pc.)

· 6 ·

SEEDS FOR A SUSTAINABILITY REVOLUTION

What Did the Renaissance Teach Us?

A fundamental NOS idea states: "Although science produces many successful outcomes, it does have its limitations." For instance, during the Scientific Revolution (1543–1686), science progressed as if it had few limits, due to the success of using the strategy of evidence-based reductionist reasoning for the first time. This success diminished Aristotle's influence on natural philosophy. The act of experimenting put the strategy of evidence-based reductionist thinking into action. Years later it resulted in accurately describing and/or explaining the phenomena of climate change.

One learns from the Scientific Revolution that it takes a radical transformation, led by highly competent leaders who will, by their accomplishments, demonstrate an expertise and perseverance in learning new ideas and demonstrating a lack of fear of replacing old ideas.

Today, climate change has been well described and adequately explained (Chapters 3–5). The issue over cleansing the atmosphere to avoid a crisis, however, has raised an excellent question. How do we control the atmosphere? In other words: how do we control what enters it? and what exits it? and how quickly? The goal is to prevent the atmosphere from overheating and taking the world through a serious tipping points (Chapter 3, section "Climate-Change Tipping Points").

Indigenous leaders think about this problem in a different way. They pay attention to how the atmosphere behaves. From the stories that *their* Elders told them, they remembered that the atmosphere holistically interacts with many geological events, all associated with the Earth and interconnected through complex webs of relationships;[1] such as: the behavior of the ocean currents, the moon's movements, the Sun's position in the sky, and the behavior of certain plants and animals at different times of the year.

However, some non-Indigenous citizens *felt* that they needed to continue to make a large profit from producing fossil fuels or mining coal. Beginning in the 1980s, they politically opposed politicians and other people who followed the advice of scientists. Their most effective "weapons" were political financial support and spreading half-truths that Canada's former Federal Minister of Environment and Climate Change labelled "climate-change cheap talk."[2]

Nero Fiddled While Rome Burned

This popular legend lacks historical truth,[3] but here it could act as a metaphor representing those climate-change-crisis deniers, some of whom had derogatory names for environmentalists. These Neros of our world today have been resistant or passive-resistant people responsible for the lack of significant progress to confront seriously the climate change crisis. A continuum of Neros include:

1. The fossil fuel legal cartels (Big Oil);
2. Industries that currently depend on fossil fuels (a few do have legitimate reasons, and those few will be accommodated in a "Sustainability Revolution)
3. Governments who do not want to give up large amounts of tax dollars from Big Oil's production of its products, and the government's tax income from the sale of those products.
4. Politicians whose election campaigns receive large legal donations from fossil fuel supporters; and

1 Cuthand (2023, Jan 7)
2 McKenna (2023)
3 History Staff (2020)

5. The public who are uncomfortable with significant changes to their lives. They tend to be motivated by a *wish* that climate change has not developed into a serious phenomenon.

This diverse group of people fiddle excuses, fiddle procrastination, fiddle hypocrisy, or fiddle gross misrepresentations; while the Earth burns hotter causing northern Canadian cities and Indigenous reserves to be evacuated and for many: loss of homes.

Indigenous Cultures on Turtle Island[4]

Chapters 5 and 6 describe the Indigenous peoples who follow an alternative path other than the Eurocentric mainstream path. For instance, Indigenous citizens will tend to interact with *Mother Earth's* atmosphere on its own terms of engagement, which requires rigorous, concrete holistic thinking. This type of thinking differs from the abstract reductionist reasoning by Western (Eurocentric) cultures.

But it has been the natural way by Indigenous cultures for over tens of thousands of years. Recall, Elders and families teach the children to use their brain, heart, body and soul in a *harmonious way*.[5] Their expertise has been proven. The evidence? A most rigorous criterion: *brute survival* under some of the most adverse circumstances, such as: extreme climates, starvation, colonization that included the horrific residential schools. Human *survival* is also the ultimate issue surrounding today's global climate-change crisis.

Chapter 6 addresses two questions: (a) Which features of the Sustainability Revolution will likely work the best at confronting the climate-change crisis? and (b) How might these features be implemented? To prepare ourselves for that challenge, here are some ideas written by two scientists, Jayalaxshmi Mistry and Andrea Berardi. They have experience at bridging both IWLN and the science of climate change. They published their ideas in the journal named *Science*,[6] as mentioned above, one of the most prestigious scientific journals for choosing its reviewers widely known for their expertise.

Yes, you certainly can trust those authors for knowing: (a) their science, and (b) the art of bridging IWLN and science. However, they are not quite up-to-date on how to write respectfully about Indigenous topics. The authors

4 Indigenous name for North America
5 Elders Little Bear, Musqua, & Scott (pc.)
6 Mistry & Berardi (2016, p. 1274)

commit most of the communication blunders discussed in Chapter 5; although it might easily have been the journal publisher's fault.

Indigenous sustainable understandings suit the uniqueness of their community's history, location and climate; along with their natural holistic ways of thinking about, and interacting with, their Mother Earth: always interacting respectfully in a reciprocal manner. It could be the essence of a community's "environmental governance."[7]

Because Indigenous understandings are always consistent with a community's norms, values, and cultural ways of using their brain, heart, body and soul, it appears to outsiders of the local community, as if the insiders' knowledge and wisdom are subjective, just because the "insiders" are acquainted with who is the source of that coming-to-know content. Western science has its own value-laden ideas because those influence how ideas are developed within a community of scientists. All knowledge is value-laden to some extent. So of course, it has a degree of subjectivity, because humans constructed it, or wrote the computer program that constructed it.[8]

The topic of Indigenous sustainable ways of living often expands into the topic of "climate justice,"[9] which is the reason behind the UN's C.O.P.'s "damage account" (Chapter 3). And of course, "Indigenous sustainable understandings" are intimately associated with environmental decision making.[10]

Sustainability Revolution Opportunities

Evidence of Indigenous communities' environmental expertise is well known. To some extent, local ecological knowledge must guide governments when they pass legislation concerning the environment. Satellite imagery is a useful tool, for example. It demonstrated that "Indigenous lands contribute substantially to maintaining healthy carbon stocks and to enhancing biodiversity; both of which include the territory adjoining those Indigenous lands."[11,12,13]

7 Mistry & Berardi (2016, website quote)
8 Mistry & Berardi (2016)
9 Solorio, Romero, & Ros (2023)
10 Solorio, Romero, & Ros (2023)
11 Mistry & Berardi (2016, p. 1275);
12 Little Bear (2011) pc.
13 Cuthand (2023, , Apr 15)

> Elder Little Bear ... speaks about science through the lens of Native American culture and teachings. He outlines several tenets of this lens and discuss how we might use them to overcome obstacles reached in the scientific community. These tenets are in constant flux, renewal, and place.

Ergo, the opportunities from a Sustainability Revolution should be *a strong collaboration between Indigenous and Western peoples in general.*[14] But in addition, any person who tends to be: quick-thinking, adaptive to two-eyed seeing, and cooperative, should be welcomed. The last criterion is suggested because limited time drives the climate-change confrontations.[15]

There is growing evidence that:[16]

> Right across Canada ... a powerful new yet quiet Indigenous movement is on the rise. It is about equity ownership, self-sufficiency and a better life. It is about reacquiring the natural resources on Indigenous territories and applying thousands of years of sustainable principles to their development.

This entrepreneur movement includes, for example: a $2.3-billion natural gas liquification and export facility; a 56% share of a $1-billion facility. They are continually instigating sustainable practices for their industries.[17]

These Indigenous investment portfolios are a vote of confidence in coordinating Indigenous IWLN and Western Science aimed at reducing the danger status of the climate-change crisis.

Mentioned earlier, but well worth repeating: an Indigenous contribution to initiating the Sustainability Revolution will come about by harmonizing the four dimensions to coming-to-knowing Mother Earth's "blanket" (the atmosphere) for cleansing: the intellectual, emotional, physical, and spiritual dimensions. This integrated Indigenous process, when translated into English, would be something like: "acting in a good way."[18]

A greater vote of confidence in the Revolution will come about by establishing a relationship with the national and provincial governments that offer climate-change research grants. This move began on a small but solid basis by Prof. Balard's work with the E.C.C.C. in Ottawa. She made the case by using her experience with IWLN. Research advances were made (Chapter 5).

14 Tsuji & Ho (2002); Mistry & Berardi (2016, Mar 16); Popp (2018); Gale (2021); Gale (2023, May)
15 Weiss, Hamann & Marsh (2013)
16 Gale & Greenspon (2021, Mar 16); Government of Canada (2023 Jan).
17 Gale & Greenspon (2021, Mar 16)
18 Pc. with about 10 Elders helping to write cross-cultural science textbooks

The liaison with governments in the future could be invaluable for helping initiate a Sustainability Revolution. The E.C.C.C. might sit at the preliminary table of organizers for the Revolution. If so, it would bring status and stability to such a table.

The Momentum Grows

In the United States at the end of 2022, the White House Office of Science and Technology Policy, and the White House Council on Environmental Quality, organized the "White House Tribal Nations Summit." The summit began by reading its mission statement which "recognized that in order to make the best scientific and policy decisions possible, the Federal Government should value and, as appropriate, respectfully include Indigenous Knowledge."[19]

> In return, a spokesperson for the American Indians politely clarified the mission statement from their perspective (perhaps to establish political equity between the two groups attending the summit).
>
> "Had our traditional cultural practices and ceremony not been outlawed and had our information keepers been listened to over the centuries, we probably would not find ourselves in the position we are today—with the losses and extinction and contamination we face as our global community.
>
> This is a valuable component of being able to face not only climate change but the preservation and protection of all of our resources."

The American Indian delegation related "evidence of success when knowledge systems were brought together;" a process they called "braiding;" as did Prof. Ballard of her climate-change research with polar bears (Chapter 5, subsection "Canadian First Nations"). The delegation explained that American Indians who were living on the land, and therefore experiencing more familial relationships with nature (Mother Earth), reported shaper observations and more accurate predictions by drawing on their four-dimensional IWLN.

The "White House Tribal Nations Summit" established a working relationship among Indigenous scholars and the White House Council on Environmental Quality.

19 Daniel, Wilhelm, Case-Scott, et al. (2022)

News Flash: The Mother Load Is at Hand

It turns out that several years ago, another U.S. major government agency, the National Science Foundation (N.S.F.), had come to the same preliminary conclusion about the potential successes of such braiding: IWLN and scientific knowledge. The N.S.F. wanted to have a more in-depth understanding, so it established the "Center for Braiding Indigenous Knowledges with Science" (C.B.I.K.S.). It has its headquarters at the University of Massachusetts Amherst, but it reaches out globally.

In the summer of 2023, C.B.I.K.S. announced a five-year international research project with a US$-30 million budget. It will involve 57 Indigenous international communities divided into eight centres: five in the U.S. and one in each of: Canada, Australia, and Aotearoa New Zealand.

The Canadian centre will be housed in the University of Alberta's Institute of Prairie and Indigenous Archaeology, aided by its Director, Kisha Supernant.[20] Alberta's focus will be on mountains and prairie locations, including voices of the Métis Indigenous Nation. The C.B.I.K.S.' general aim is to confront climate change from the point of view of making policy decisions that harmonize with Indigenous leaders; in other words: decisions for survival during the next seven generations, *beginning immediately*. This requires the understandings of our environmental change and how it affects our food and cultural systems on local and global scales.

The methodology most likely will entail: (a) braiding IWLN with related Western science; (b) and then examining a diversity of results by listening to the stories of many peoples' experiences: those who participated in the braiding, and those those who observed as outsiders. Further details were not finalized when this book went to print.

Simply put, *coincidentally*, what the N.S.F. is proposing to accomplish with its $30-million investment *is to investigate the feasibility of this book's proposal* for Canadians to initiate a worldwide Sustainability Revolution. That is the major reason that makes this book both *unique and current.*

C.B.I.K.S. (Center for Braiding Indigenous Knowledge with Science) spelled out an extensive and sophisticated research program, the results of which during its five-year duration will help Canada become seriously organized into launching a Sustainability Revolution, in conjunction with cooperating global countries. With it, the torch is being passed to the younger generation.

20 Brown (2023, Sep 12)

By itself, C.B.I.K.S. investment clearly indicates support for Canada making a high priority for a multi-science effort (i,e., including some social sciences such as economics and political science) to invest in a host of coordinated research and development proposals focussed on ending as quickly as possible, the current crisis caused by a neglect or resistance to climate change since 1988.

Multi-science-Indigenous collaborations could be launched in the near future. At the local R.&D. sites, a three- to five-day cultural immersion on a reserve would the first step, followed by paced-out sessions on the local Indigenous language (a window into the group's worldview), emphasizing the structure of their Indigenous language. To reach first base is to be able to simultaneously activate one's personal combination of their brain, heart, body and soul. This is not theory; it arises from hundreds of thousands of years of humans developing this survival skill. The climate-change crisis is humanity's test.

The Persuasive Power of Angry Teenagers

There are some thoughtful generalities that show up in climate change discussions. On the one hand, "The way we see the world shapes the way we treat it."[21] On the other hand, there is the famous quote: "We see the world, not as it is, but as we are; or as we are conditioned to see it."[22] The two quotes are interrelated.

Their relationship would suggest that a Sustainability Revolution must focus on *persuading* those holding onto their pure-profit-society world. Thus, given the seriousness of climate change's crisis, no-holds-barred persuasion strategies need to be seriously considered, instead of applying psychological "conditioning" theories. Wallace-Wells' book, *The Uninhabitable Earth: Life After Warming* uses persuasion cleverly.

Talented persuaders, including professional marketers, need to be recruited into the public square as, daughter of Dr. David Suzuki, spokespersons for the next generation's youth. They could play a critical role in getting important support for the Revolution.

Severn Cullis-Suzuki was just 12-years-old when, at the Rio de Janeiro Earth Summit in 1992, she made an impassioned plea to avert an ecological disaster. She spoke truth to power. The powerful audience reacted instantly. A deafening silence filled the large Rio assembly room. Her 1992 story continues to

21 Suzuki (2015) as quoted in Goodreads (2023)

22 Covey (n.d., p. 129)

be retold today; for example, by the 2019 Victoria, BC, *TimesColonist* headlines: "Before Greta, there was Severn, the girl who silenced the world."[23]

Who is Greta? At the age of 16, a smart and angry Swedish Greta Thunberg spoke at the 2019 United Nations Climate Action Summit at U.N. headquarters in New York. Her address began this way:[24]

> My message is that we'll be watching you.
>
> This is all wrong. I shouldn't be up here. I should be back in school on the other side of the ocean. Yet you all come to us young people for hope. How dare you!
>
> You have stolen my dreams and my childhood with your empty words. And yet I'm one of the lucky ones. People are suffering. People are dying. Entire ecosystems are collapsing. We are in the beginning of a mass extinction, and all you can talk about is money and fairy tales of eternal economic growth. How dare you!

She, too, silenced the crowd.

Although the world listened in each case, little change at these international meetings has taken place (e.g., C.O.P. meetings). The reasons were detailed in Chapter 3. But the point to be taken here is that the power of persuasion by angry articulate teenagers is real, and should not be overlooked.

The Persuasive Power of Litigation

The fossil fuel industries (i.e., coal, oil, and gas) say they are planning to continue to produce their products, albeit at a gradually slower pace over time. At their usual pace, the Earth will most likely slip past a serious tipping point of climate change, then eventually into a fiery oblivion.[25] "The rights to life, equality and security of the person will be meaningless on a dying planet and therefore must be interpreted with reference to ecological sustainability."[26]

1. "So big emitters like fossil fuel companies absolutely they need to be aware of their potential exposure . . . in terms of potential liability."[27]
2. "A recent Ontario court case, Mathur v Ontario, was the first Canadian children's climate case to be decided on the merits."[28] Although

23 Crawford (2019)
24 NPR Staff (2019)
25 Guterres (2022)
26 Wood (2023)
27 Shingler (2023)
28 Wood (2023)

the Ontario Superior Court of Justice dismissed the lawsuit in April of 2023, "The case raises important issues of justiciability, standing, constitutional rights and climate science."[29] There are good reasons that it will be appealed all the way to the Supreme Court.[30]

3. "Seven Saskatchewan residents between the ages of 15 and 80 claim the government's action to expand gas-fired electricity generation violates their Charter rights."[31]
4. Indigenous Canadians tend to be more vulnerable to the effects of climate change. "It's our right to use our voice in court to fight for our future, to assert our rights to a safe and healthy environment," said Wabegijig, an Anishinaabe activist and Caribou Clan from Timiskaming First Nation.[32]

These court cases occur due to the fear, frustration and desperation of angry youth and their advocates.[33] They realize that Big Oil has infiltrated government circles through lobbying, and has controlled the outcomes of international meetings dedicated to solving the climate-change crisis (Chapter 3). "After a century of wielding extraordinary economic and political power, America's petroleum giants face a reckoning for driving the greatest existential threat of our lifetimes."[34]

"An unprecedented wave of lawsuits, filed by cities and states across the U.S. aim to hold the oil and gas industry to account for the environmental devastation caused by fossil fuels—and covering up what they knew along the way."[35] In other words, the community is insisting that the base-ball team pay for the window broken by an errant fly ball out of the neighborhood park.

In August 2023, a group of Montana youths successfully sued their state legislature to cease "continued development of fossil fuels."[36] District Court Judge Kathy Seeley ruled that Montana's continued development violates a clause in its state constitution that guarantees its citizens the right to a "clean and healthful environment."

29 Wood (2023)
30 Wood (2023)
31 Shingler (2023)
32 Shingler (2023)
33 McGreal (2021)
34 McGreal (2021)
35 McGreal (2021)
36 Nilsen (2023)

In Canada, the David Suzuki Foundation has been "supporting 15 youth from seven provinces and a territory taking the Canadian government to court for violating their charter rights by perpetuating climate change."[37]

In the Netherlands, a successful 2015 court challenge by hundreds of citizens forced their government to reduce greenhouse gas emissions more than their new policy had stated. In 2021, "a Dutch court ordered Royal Dutch Shell to cut its carbon emissions. That decision has been appealed."[38]

Perhaps the most promising organized litigation is the west-coast grassroots movement initiated by Avi Lewis[39] entitled "Sue Big Oil." Avi is a documentary filmmaker, climate activist, and university educator. One of his films, *This Changes Everything,* was featured at the 2015 COP_{21} meeting in Paris.[40]

> For decades, oil and gas corporations have known that burning fossil fuels would cause the heat waves, wildfires, drought and flooding that we're now experiencing in BC. . . .These multinational companies spent millions to deceive, deny and distract us on their way to billions in profit by preventing action on climate change. . . .Fossil Fuel giants have captured governments and banks, extracting mega-profits while the world burns. It's time to Sue Big Oil.

Four of Avi's several distinguishing features are:

1. He is currently a not-yet-elected political candidate for the national New Democratic Party (NDP) for a south BC riding. So he has a chance to be a member of the Canadian Parliament if he wins in the next federal election; expected in 2025.[41]
2. His "Sue Big Oil" movement currently involves small local governments across British Columbia.
3. He is an associate professor at the University of British Columbia's Centre for Climate Justice.[42]
4. He was born in Toronto into a well-known political family: (a) his father, Stephen Lewis, and his grandfather, David Lewis, were both highly active in provincial and federal politics.

37 David Suzuki Foundation (2023)
38 Shingler (2023)
39 Lewis (n.d.-2)
40 Lewis (n.d.-1)
41 The Canadian Press (2021)
42 Lewis (n.d.-1)

Climate litigation against governments and corporations is likely to become more common as the science of climate change and the grounds for a legal challenge become clearer.

Climate litigation could become a power strategy for establishing or supporting a Sustainability Revolution.[43]

A Canadian leader from each side of the climate-change issue—a strong-willed hero environmentalist (Dr. David Suzuki) and a well-known avid business man (Mr. Jim Pattison)—argue their case in this book's Epilogue. It is personal and it portrays the real world for you to sort out for yourself.

43 Shingler (2023)

EPILOGUE: AN ECONOMIC/CULTURAL TUG-OF-WAR

In February 2021, an unusually powerful snow storm knocked out most of the electrical power across the State of Texas. A month later, reporter Gabriel Friedman from the Financial Post[1] interviewed British Columbia billionaire, Jim Pattison, who had been an avid climate change denier at one time.

Over a period of several days, Gabriel and Jim spoke at long distance about Jim's personal journey in understanding climate change. Gabriel also spoke privately with David Suzuki on the same topic after each conversation with Jim. Gabriel wrote his article as a discussion between Jim and David with Gabriel acting as the moderator. The remaining citations in this Epilogue come from the Financial Post article composed by Gabriel, unless otherwise stated.

Pattison's business empire includes car dealerships, a major grocery store chain (Save-on-Foods), and investments in commercial fishing, forestry, coal mining, and many others. At the same time, his philanthropy is well known, highlighted by the 2019 opening of the Jim Pattison Children's Hospital in Saskatoon, Saskatchewan. It is a $286 million new hospital to which about $75 million was donated by Pattison.

1 Friedman (2021, website quotes)

As mentioned, Gabriel Friedman also spoke with the genetics scientist, Dr. David Suzuki, known for his environmental activism that included his 1980s prophetic public warnings about the impending effects of climate change. Suzuki became very unpopular among corporate Canada.

Surprisingly, Pattison and Suzuki had developed an amicable relationship when they sailed up the Vancouver coast together on several occasions while both independently investigated the fishing industry based on two different agendas. At the same time, they publicly and forcefully opposed each other on matters related to the environment, especially climate change.

But over time, Pattison seriously began to consider Suzuki's arguments about climate change. At first, however, Pattison had thought, "this guy was really off base."[2] But later he confessed, "guys like me were off base." Pattison went on to say, "I thought he was a radical and today (it looks like) he was a prophet."

The importance of climate change had dramatically evolved for Pattison over "two or three years"[3] prior to his interview with Friedman. That could also have happened, however, when Pattison took a big financial loss on his investments in his Texas' electricity companies, due to that massive snow storm.

Sometimes it takes an emotional push to help some people understand:

1. just how real and devastating climate change can be; and
2. that equally drastic reaction is needed.

Other people are influenced by photographs, or by victims' emotional statements on video.

"At one time we never considered climate change,"[4] Pattison mentioned. "Today, it's the number one thing we think about when we talk about a new investment."

Suzuki later responded to Gabriel that changing one's mind on such a major issue isn't as simple as a sudden overnight conversion. He said he thinks Pattison accepted the science of climate change years ago, but was too focused on making a profit from his businesses to take the serious actions required to stop climate change from ripping the planet apart.

"He knows his investment in coal is a problem," said Suzuki. (Burning coal is the dirtiest atmospheric carbon polluter there is.) Pattison had explained to Suzuki that his company has a 40-year legal contract to supply coal (i.e., to a

2 Friedman (2021, website quotes)
3 Friedman (2021, website quote)
4 Friedman (2021, website quotes)

Texas corporation that produces electricity by burning it). He would have to sell all of his coal assets at a loss. This contradicted a pure-profit society's business ideology.

Suzuki told the reporter that Pattison knows that we don't have 40 years. We only have about a decade or two to avoid the really catastrophic climate events. Those costs will dwarf Pattison's costs for breaking his contract.

The tug-of-war is experienced differently depending on the context. If it's an economic question (e.g., Should Pattison take a hit for team humanity by selling off his coal assets at a loss?). If so, he would find himself in a tug-of-war with peer pressures from both sides of the issue. Knowing the right answer is one thing, acting on it is something else. In other words, knowledge versus wisdom.

Although Suzuki told Friedman that Pattison "was too focussed"[5] on his businesses to make major changes to his financial portfolio, Suzuki's phrase, "too focussed," seems to be a nebulous tautology. Should we see the issue as an ethical one? In other words, does Pattison know what he *should do* but does not have the fortitude to do it? His generous philanthropic record strongly dismisses that ethics issue.

Or is Pattison as addicted to his pure-profit society's ideology, as Suzuki is addicted to his scientific subculture? "Addiction" connotes a negativity that raises more questions than suitable solutions.

Over the years, both Suzuki and Pattison individually rationalized contributing to two projects belonging to each other; for instance, West Coast fishing operations. "Ultimately," Suzuki told Friedman, "Pattison told him he needed to end the contributions because [Pattison] was feeling pressure from peers that he was supporting environmental activism."[6]

"Even today, if they agree that climate change is real, the same tension on what to be done about it remains," Suzuki continued.

"The economy has completely ignored what nature does to keep the planet inhabitable,"[7] said Suzuki. "This is the kind of thing business people have a hard time absorbing because it means they can't go on doing business as they have been."

Networks of legal business contracts, similar to Pattison's coal mining contracts, seem to act like gigantic impenetrable spider webs trapping people from making quick progress towards net-zero. And as mentioned above,

5 Friedman (2021, website quotes)
6 Friedman (2021, website quotes)
7 Friedman (2021, website quotes)

those contracts will be the first to disintegrate during a massive, non-reversable drought; brought on, for instance, by a massive heat wave.

"[Pattison] said his companies are 'changing gears as fast as practical to try to be more environmentally involved'."[8] This type of comment lacks specificity, similar to many of the country delegates at the COP meetings.

"Our people including me didn't put that much confidence in what David Suzuki was talking about," Pattison confessed. "But he was right."[9]

In his column, Friedman mentioned some encouraging news for Mark Carney: "The change in Pattison's views on climate change mirrors a broader shift in the corporate world."[10] For instance, the chief executive of a major investment firm, BlackRock Inc., wrote: "investors in mutual funds and exchange traded funds had invested US$ 288 billion globally in sustainable assets in 2020—a 96 per cent increase from 2019."

The Road Ahead

A scientific explanation from anthropology offers additional clarity to the economic and political perspectives that help explain the tug-of-war between the David Suzukis and the Jim Pattisons of the world. Eminent anthropologist Edward T. Hall[11] turned his attention to the mathematical beliefs established by Plato's *purity-of-the-mind ideology* (i.e., whatever comes out of Plato's mind is the truth).

Edward T. Hall, wrote that Plato's philosophical concept, "purity of the mind," was actually an intellectual mirage: "*What has been thought of as the mind is actually internalized culture*" (original emphasis).[12]

Some ideas originally thought to have had their origins in a human's mind actually have their origins in a group's culture. This could be an analogy for Suzuki's and Pattison's tug-of-war over the proper response to the climate change crisis. First of all, climate change on the surface is a scientific or engineering problem to solve. But understanding climate change at a deeper level, the fundamental problem to solve is a *cultural* problem; the culture of our pure-profit society that reveres financial gains.

8 Friedman (2021, website quote)
9 Friedman (2021, website quotes)
10 Friedman (2021, website quotes)
11 Hall (1976)
12 Hall (1976, p. 192)

Our current predominant pure-profit society's extreme worldview needs to undergo a revolutionary change *from* its absolutist position, *to* a worldview that embraces a noticeable degree of harmony among people and their relationship with nature (i.e., sustainability) to a significant degree. For many, this means updating:

- Our *assumptions* about the physical world,
- *How* we know what we know, and
- Our *values, beliefs, and ideologies.*

In other words, updating society to harmonize with the emerging Sustainability Revolution means that a large majority of people need to make this transition in order to save humanity. This corresponds to Kepler's transition from an ancient Greek worldview to his modern worldview that led him to understand the solar system. For today's Sustainability Revolution, what does "to make this transition" actually look like? It is not just an update; it is an historic worldwide change of culture.

To Build Strength for the Side That Is Right

Pattison confessed, "He [Suzuki] was right." The way Pattison worded his confession reveals an assumption he was making: the issue is one of "right" versus "wrong." He was thinking with a subliminal dichotomy. Dichotomies dictate an all-or-nothing strategy when making a transition. But in the context of moving into the Sustainability Revolution, Pattison made a false assumption. A dichotomy is not the only strategy available. A continuum strategy is available. It goes from one extreme to the other extreme, but there is lots of room on either side of its middle position. A continuum is NOT an all-or-nothing strategy that makes some people feel that a change is risky. They tend to resist risks.

Yes, the Sustainability Revolution needs some people at the extreme endpoint of sustainability. You have heard about some them in this book. But there is plenty of room on the sustainability side of the continuum for significant supporters who make a difference when in large numbers. Initiating lawsuits against Big Oil is not a requirement for joining the sustainability side, but moral support for them in your own way is appreciated.

If you are asking yourself what specifically you can do to plant your feet on the sustainable half of the continuum, one place to begin is to draw on the David Suzuki Foundation's menu of 10 practical actions that move you further *away from* the pure-profit society side of the continuum, to joining a sustainable society with reasonable profits side of the continuum model.

Ten top suggestions are detailed in the Foundation's website, free of change.[13] Consider them as 10 places along one side of your personal sustainability continuum. Also, most provinces in Canada have active environment organizations. They can help you move along your continuum. The U.S. Environmental Protection Agency (E.P.A.) also publishes practical ideas on how to live more sustainably.[14]

Obvious to readers of this book, my last suggestion is to arm yourself with a Catherine McKenna climate-change-cheap-talk detector.[15]

The Suzuki Foundation actively advocates for, and/or sponsors, projects that:

- *Respect, restore and protect forests, wetlands, grasslands and oceans.* Preserving ecosystems supports species and averts the worst effects of climate change.
- *Work in partnership with Indigenous Peoples.* Recognizing their rights and incorporating their wisdom (i.e., living in a good way). They have gained this wisdom from surviving in a particular place for millennia.
- *Advance Canada's climate ambitions.* The Suzuki Foundation's research shows how to get to an affordable, reliable 100% zero-emissions national electricity grid in just 12 years.
- *Ban fossil fuel development.* Quebec is Earth's first jurisdiction to end oil, gas and coal extraction in its territory. Arm yourself with More jurisdictions must follow.

Climate litigation against governments and corporations is likely to become more common as the science of climate change and the grounds for a legal challenge become clearer.

13 David Suzuki Foundation (2022; pc.)

14 E.P.A. (2022)

15 McKenna (2023)

The Final Thought

This book's true story is yours. "Do with it what you will. Cry over it. Get angry. Forget it. But don't say in the years to come that you would have lived your life differently if only you had read this story. You've read it now."[16]

16 King (2003, p. 119)—Indigenous author of "*The Truth about Stories*."

BIBLIOGRAPHY

12:eleven. (2023, May 1). The importance of ESG in the oil and gas industry. Author. <https://www.12eleven.com/news/the-importance-of-esg-in-the-oil-and-gas-industry>

Afzal, M., Victor, D. G., & Carlton. E. (2022, Nov 12). Around the halls: What's happening at COP_{27} in Egypt? Brookings Education. <https://www.brookings.edu/blog/planetpolicy/2022/11/10/around-the-halls-whats-happening-at-cop27-in-egypt/>

Aikenhead, G. S. (1979). Science: A way of knowing. *The Science Teacher*, *46*(September), 23–25.

Aikenhead, G. S. (1996). Science education: Border crossing into the subculture of science. *Studies in Science Education*, *27*, 1–52.

Aikenhead, G. S. (1997). Toward a First Nations cross-cultural science and technology curriculum. *Science Education*, *81*(2), 217–238.

Aikenhead, G. S. (Ed.). (2000). *Rekindling traditions: Cross-cultural science & technology units* project. <https://education.usask.ca/ccstu/>

Aikenhead, G. S. (2002). *Rekindling traditions* (website). *Canadian Journal of Science, Mathematics, Technology, and Education*, *2*(2), 287–304.

Aikenhead, G. S. (2006a). *Science education for everyday life: Evidence-based practice*. Teachers College Press, Columbia University.

Aikenhead, G. S. (2006b). Towards decolonizing the pan-Canadian science framework. *Canadian Journal of Science, Mathematics and Technology Education*, 6(4), 387–399.

Aikenhead, G. S. (2007). Humanistic perspectives in the science curriculum. In S. K. Abell & N. G. Lederman (Eds.), *Handbook of research on science education* (pp. 881–910). Lawrence Erlbaum Associates.

Aikenhead, G. S. (2008). Objectivity: The opiate of the academic? *Cultural Studies of Science Education, 3*, 581–585.

Aikenhead, G. S. (2017). Enhancing school mathematics culturally: A path of reconciliation. *Canadian Journal of Science, Mathematics and Technology Education, 17*(2: Special Monograph Issue), 73–140.

Aikenhead, G. S. (2020). School science and mathematics storylines. *Canadian Journal of Science, Mathematics and Technology Education, 20*(4), 682–699.

Aikenhead, G. S. (2023). Humanistic school science: Research, policy, politics and classrooms. *Science Education, 107*(2), 237–260. <https://doi.org/10.1002/sce.21774>

Aikenhead, G. S., Brokofsky, J., Clark, C., et al. (2014). *Enhancing school science with Indigenous knowledge: What we know from teachers and research.* Saskatoon Public School Division, and Amazon, CA.

Aikenhead, G. S., & Elliott, D. (2010). An emerging decolonized science education in Canada. *Canadian Journal of Science, Mathematics and Technology Education, 10*(4), 321–338.

Aikenhead, G., & Michell, H. (2011). *Bridging cultures: Indigenous and scientific ways of knowing nature.* Pearson Canada.

Aikenhead, G. S., & Ogawa, M. (2007). Indigenous knowledge and science revisited. *Cultural Studies of Science Education, 2*(3), 539–591.

Amazon.ca. (n.d.). The Deadly Rise of Anti-science: A Scientist's Warning. <https://www.amazon.ca/Deadly-Rise-Anti-science-Scientists-Warning/dp/1421447223>

Ayed, N. (2021, Oct 22). The danger of COVID-19 misinformation is mind-boggling (Transcript). Canadian Broadcasting Corporation's program IDEAS. <https://www.cbc.ca/radio/ideas/the-danger-of-covid-19-misinformation-is-mind-boggling-says-dr-anthony-fauci-1.6218070>

Ayed, N. (2022, Dec 19). Nowhere left to run: Climate reparations (podcast). Canadian Broadcasting Corporation's program IDEAS. <https://www.cbc.ca/listen/live-radio/1-23-ideas/clip/15955840-nowhere-left-run-climate-reparations>

Bang, M., & Medin, D. (2010). Cultural processes in science education: Supporting the navigation of multiple epistemologies. *Science Education*, 94(6), 1008–1026. <https://onlinelibrary.wiley.com/doi/full/10.1002/sce.20392>

Barnhardt, R., & Kawagley, A. O. (2005). Indigenous knowledge systems and Alaska Native ways of knowing. *Anthropology and Education Quarterly*, 36, 8–23.

Barwell, R. (2013). The mathematical formatting of climate change: Critical mathematics education and post-normal science. *Research in Mathematics Education, 15*(1), 1–16.

Battiste, M. (1986). *Micmac literacy and cognitive assimilation.* UBC Press.

Battiste, M. (Ed.). (2000). *Reclaiming Indigenous voice and vision.* UBC Press.

Battiste, M. (2013). *Decolonizing education: Nourishing the learning spark.* Purich Press.

Begay, D. H., Maryboy, N. C., et al. (2021). Two eyed seeing: Navajo (Diné) astronomy & NASA Moon to Mars (audio tape of live performance). Native Skywatchers. <https://www.nativeskywatchers.com/two-eyed-nasa-navajo.html>

Bigham, B. (2021, July 28). At this point in the pandemic, it's misinformation that's killing people. *The Globe and Mail.* <https://www.theglobeandmail.com/opinion/article-at-this-point-in-the-pandemic-its-misinformation-thats-killing-people/>

Board, J. (2022, Nov 3). 7 reasons why Egypt's climate conference matters. *Channel New Asia.* <https://www.channelnewsasia.com/sustainability/cop27-egypt-climate-change-conference-why-it-matters-3028416?cid=cna_flip_070214>

Borovik, A. V. (2017). Mathematics for makers and mathematics for users. In B. Sriraman (Ed.), *Humanizing mathematics and its philosophy* (pp. 309–327). Springer International Publishing.

Bouchard, L. (2023a, Jul 11). The warning. CBC Ideas. Suzuki's survival guide. <https://www.cbc.ca/radio/ideas/david-suzuki-s-survival-guide-a-retrospective-1.6894102>

Bouchard, L. (2023b, Jul 18). How we got to this point (podcast).

Brown, M. (2023, Sep 12). U of A archeologist tapped to help lead major exploration of Indigenous knowledge. University of Alberta Folio. <https://www.ualberta.ca/folio/2023/09/u-of-a-archeologist-tapped-to-help-lead-major-exploration-of-indigenous-knowledge.html>

CA Finance Yahoo. (2023, Mar 20). A survival guide for humanity. Author. <https://ca.finance.yahoo.com/video/u-n-chief-urges-faster-170436299.html>

Cajete, G. (1999a). *Igniting the sparkle: An Indigenous science education model.* Skyand, NC: Kivaki Press.

Cajete, G. (1999b). *A people's ecology: Explorations in sustainable living.* Clear Light Publishers. ISBN 13: 9781574160284

Cajete, G. (2000). *Native science: Natural laws of interdependence.* Clear Light Publishers. <https://www.youtube.com/watch?v=3BqoZhp2Zn4>

Cajete, G. (2015). *Indigenous community: Rekindling the teachings of the seventh fire.* Living Justice Press. ISBN 978-1-937141-17-2

Cajete, G. (n.d.). Explore from here. A brief history of quantum mechanics. *YouTube.* BanffEvent.

Canada Energy Regulator (C.E.R.). (2023, Jun 20). Canada's energy future 2023: Energy supply and demand projections to 2050 (news release) (website quote). Author. <https://www.cer-rec.gc.ca/en/about/news-room/news-releases/2023/canada-energy-regulator-projects-major-transformation-canada-energy-system-net-zero-world.html>

Carbonbrief. (2020). Explainer: Nine "tipping points" that could be triggered by climate change. [map] <https://www.carbonbrief.org/explainer-nine-tipping-points-that-could-be-triggered-by-climate-change/>

Card, D., & Payne, A. A. (2017, September). High school choices and the gender gap in STEM (Working Paper 23769). Cambridge, MA: National Bureau of Economic Research. <https://www.nber.org/papers/w23769>

Carney, M. (2020a, December 16). Lecture 1: From moral to market sentiments (transcript). BBC Reith Lectures 2020: How we get what we value. <https://www.bbc.co.uk/programmes/m000py8t>

Carney, M. (2020b, December 23). Lecture 4: From climate crisis to real prosperity (transcript). BBC Reith Lectures 2020: How we get what we value. www.bbc.co.uk/radio4

Carney, M. (2020c). BBC Reith Lectures 2020 (Utube version). <https://www.youtube.com/watch?v=uvw-aC0KLD4>

Carney, M. (2021). *Value(s): Building a better world for all.* Penguin Random House Canada.

Carney, M. (n.d.). Co-Chair, GFANZ, UN special envoy for climate action and finance. Author. <https://www.gfanzero.com/about/mark-carney/>

Caspar, M. (1993). *Kepler.* Dover Publication, Inc.

CBC. (2015). 5 key points in Paris 2015. Author. <https://www.cbc.ca/news/world/paris-agreement-key-climate-points-1.3362500>

CBC Ideas. (2023, Jul 11). David Suzuki's survival guide: A retrospective. Author. <https://www.cbc.ca/radio/ideas/david-suzuki-s-survival-guide-a-retrospective-1.6894102>

CBC News. (2019, Apr 4). Canada warming at twice the global rate, leaked report finds. Author. <https://www.cbc.ca/news/science/canada-warming-at-twice-the-global-rate-leaked-report-finds-1.5079765>

CBC News. (2022, Sept 23). Facing calls to resign, World Bank president changes answer on climate crisis. Author. <https://www.cbc.ca/news/world/malpass-world-bank-climate-change-1.6592799>

CBC News (2023, Aug 12). Maui wildfires displace thousands of people, cause scramble for shelter. *CBC News.* <https://www.cbc.ca/news/world/hawaii-maui-wildfires-death-toll-1.6934959>

C.B.I.K.S. (Center for Braiding Indigenous Knowledges and Science). (n.d.) (website). <https://www.umass.edu/gateway/research/indigenous-knowledges>

CCL (Canadian Climate Institute). (2022). 440 Megatonnes: Tracking Canada's climate policy progress. Author. <https://440megatonnes.ca/?gclid=EAIaIQobChMI9MzwvsPn-wIVaSytBh0cXQ7SEAAYBCAAEgKhffD_BwE>

C.E.R.. (2023). See Canada Energy Regulator (2023).

Chung, E. (2023, Jul 17). Imminent collapse of Atlantic current could lead to sudden, irreversible climate impacts, new study suggests. *CBC News.* <https://www.cbc.ca/news/science/climate-tipping-points-1.6918795>

ClientEarth Communications. (2020, Dec 22). What is a carbon sink? Author. <https://www.clientearth.org/latest/latest-updates/stories/what-is-a-carbon-sink/>

Cobern, W. W., Adams, B. A. J., Pleasants, B. A.-S., et al. (2022). Do we have a trust problem? Exploring undergraduate student views on the tentativeness and trustworthiness of science. *Science & Education* (online January). <https://doi.org/10.1007/s11191-021-00292-1>

Cohen, I. B. (1960). *Birth of a New Physics.* Penguin Random House.

Cohen, I. B. (1970). History of science lecture. Personal communication. Harvard University.

Conant, J. B. (1947). On understanding science: An historical approach. *American Scientist, 35*(1), 33–55.

Covey, S. R. (n.d.). We see the world, not as it is. Quote Fancy. <https://quotefancy.com/quote/909276/Stephen-R-Covey-We-see-the-world-not-as-it-is-but-as-we-are-or-as-we-are-conditioned-to>

Crawford, T. (2019, Sep 27). Before Greta, there was Severn, the B.C. girl who silenced the world (website). *TimesColonist.* <https://www.timescolonist.com/life/before-greta-there-was-severn-the-bc-girl-who-silenced-the-world-4675949>

CTV News. (2023, Jul 23). Two children among four missing after record-breaking Nova Scotia downpours. Author. <https://atlantic.ctvnews.ca/two-children-among-four-missing-after-record-breaking-nova-scotia-downpours-1.6490336>

Cuthand, D. (2023, Jan 7). Indigenous voices useful in the move to a greener future. *Saskatoon StarPhoenix*, A4. <https://thestarphoenix.com/opinion/columnists/doug-cuthand-indigenous-voices-useful-in-the-move-to-a-greener-future>

Daigle, K. et al. (2022). First draft of climate deal at the COP_{27} summit. <https://www.theglobeandmail.com/business/international-business/article-first-draft-of-climate-deal-at-cop27-summit-retains-15c-limit-but-many/>

Daniel, R. A., Wilhelm, T. A., Case-Scott, H., et al. (2022, Dec 2). What is "Indigenous Knowledge" and why does it matter? Integrating ancestral wisdom and approaches into federal decision-making. The White House. <https://www.whitehouse.gov/ostp/news-updates/2022/12/02/what-is-indigenous-knowledge-and-why-does-it-matter-integrating-ancestral-wisdom-and-approaches-into-federal-decision-making/>

David Suzuki Foundation. (2022). Top 10 things you can do about climate change. Author. <https://davidsuzuki.org/what-you-can-do/top-10-ways-can-stop-climate-change/>

David Suzuki Foundation. (2023). Youth climate lawsuit (website). Author. <https://davidsuzuki.org/project/youth-climate-lawsuit/>

Davies, A., & Cursino, M. (2022, July, 20). Wildfires rage in Greece, Spain and Italy as heatwave moves across Europe. *BBC News*. <https://www.bbc.com/news/world-europe-62235561>

Davis, W. (2009). *The Wayfinders: Why ancient wisdom matters in the modern world*. House of Anansi Press.

Dion, S. D. (2009) *Braiding histories: Learning from Aboriginal peoples' experiences and perspectives*. UBC Press.

Down to Earth. (2022). COP_{27}. Author. <https://www.downtoearth.org.in/climate-change/coplist>

Duignan, B. (n.d.). Dunning-Kruger effect. *Encyclopedia Britannica*. <https://www.britannica.com/science/Dunning-Kruger-effect>

Duran, P. H. (2013). *The condor and the eagle: Uniting heart and mind in search of a new science worldview*. Eaglehouse Publications.

Duschl, R. A. (2022). Naturalizing the nature of science: A response to and acknowledgement of Klopfer & Aikenhead. *Science Education, 106*(3), 484–489.

Editorial Board. (2023, Oct 19). Provinces have climate powers—and duties. *Globe and Mail* (editorial). <https://www.theglobeandmail.com/opinion/editorials/article-provinces-have-climate-powers-and-duties/>

E.N.R. (Environment and Natural Resources). (n.d.). U.N. conference on climate change: COP_{27} in Egypt. <https://www.canada.ca/en/services/environment/weather/climatechange/canada-international-action/un-climate-change-conference/cop27-summit.html>

Environmental Century Files. (2015). Graphical summary. *Wordpress*. <https://environmentalcentury.files.wordpress.com/2015/12/graphical-summary-1024x785-638x489.jpg>

E.P.A. (U.S. Environmental Protection Agency). (2022). Learn about sustainability. <https://www.epa.gov/sustainability/learn-about-sustainability>

Ermine, W. J. (1995). Aboriginal epistemology. In M. Battiste & J. Barman (Eds.), *First Nations education in Canada: The circle unfolds* (pp. 101–112). University of British Columbia Press.

Eurostat. (2022, Oct 24). Losses from climate change: €145 billion [CAN$ 212 billion] in a decade. Author. <https://ec.europa.eu/eurostat/web/products-eurostat-news/-/ddn-20221024-1>

Feinstein, C. (2023, Jun 28). Canada is ill-prepared for "disruption" in oil and gas sector as global demand is projected to drop. *Toronto Star.* <https://www.thestar.com/business/canada-is-ill-prepared-for-disruption-in-oil-and-gas-sector-as-global-demand-is/article_4d9c3ce5-1185-514d-8278-c28f168078e1.html>

Fenech, A. (2013, Jul 2). Reflections on the Toronto conference—25 years later. UPEI Climate Lab, University of Prince Edward Island. <https://projects.upei.ca/climate/2013/07/02/reflections-on-the-toronto-conference-25-years-later/>

Frederick, W. A. (1991). Science and technology education: An engineer's perspective. In S. K. Majumdar, L. M. Rosenfeld, P. A. Rubba et al. (Eds.), *Science education in the United States: Issues, crises and priorities* (pp. 386–393). The Pennsylvania Academy of Science.

Friedman, G. (2021, Mar 3). The activist and the billionaire: How David Suzuki changed Jim Pattison's mind on climate change. *Financial Post.* <https://financialpost.com/commodities/energy/the-activist-and-the-billionaire-how-david-suzuki-changed-jim-pattisons-mind-on-climate-change>

Funtowicz, S. O., & Ravetz, J. R. (1993). Science for the post-normal age. *Futures, 25*(7), 739–755.

Gale, S. (2023, May) Involve Indigenous Peoples. *The Globe and Mail Magazine.* p. 36. <https://www.theglobeandmail.com/business/rob-magazine/>

Gale, S., & Greenspon, E. (2021 Mar 16). First Nations are not obstacles but an opportunity to get things done in just and sustainable ways. *The Globe and Mail.* <https://www.theglobeandmail.com/business/commentary/article-first-nations-are-not-obstacles-but-an-opportunity-to-get-things-done/?login=true>

Ganti, A. (2022, May 31). What is asset management? *Invesgopedia.* <https://www.investopedia.com/terms/a/assetmanagement.asp>

Garroutte, E. M. (1999). American Indian science education. The second step. *American Indian Culture and Research Journal, 23*(4), 91–114. <https://meridian.allenpress.com/aicrj/article-abstract/23/4/91/210717/American-Indian-Science-Education-The-Second-Step?redirectedFrom=PDF>

Gatti, L. V., Basso, L. S., Miller, J. B., et al. (2021). Amazonia as a carbon source linked to deforestation and climate change. *Nature*, 595, 388–393. <https://doi.org/10.1038/s41586-021-03629-6>

Globe and Mail. (2022, Aug 25). Oil companies need a push on climate (Editorial). Author. p. A10 https://www.theglobeandmail.com/opinion/editorials/article-mapping-the-future-of-canadian-oil-in-a-net-zero-world/

Globe and Mail. (2022, Aug 25). Oil companies need a push on climate (Editorial). *The Globe and Mail.* p. A10. https://www.theglobeandmail.com/opinion/editorials/article-oil-companies-need-to-be-pus hed-on-clim ate/?login=true

Globe and Mail. (2022, Dec 6). Canadian oil in a net zero world (Editorial). *The Globe and Mail* (Prairie Edition). <https://www.theglobeandmail.com/opinion/editorials/article-mapping-the-future-of-canadian-oil-in-a-net-zero-world/>

Goodreads. (2023). Quotes (Goodreads.com website). <https://www.goodreads.com/quotes/7272948>

Google searches (Aug, 2023):

top 4 countries producing solar energy—Google Search

top 4 countries producing wind energy—Google Search

top country producing hydrogen energy—Google Search

top 4 countries producing battery energy—Google Search

Gore, A. (2007). *An inconvenient truth: The crisis of global warming.* Penguin Publishing Group.

Government of Canada. (2022, Mar). Canada's climate plans and targets. Author. <https://www.canada.ca/en/services/environment/weather/climatechange/climate-plan/climate-plan-overview.html>

Government of Canada. (2023). Nature-based climate solutions (website). <https://www.canada.ca/en/services/environment/our-environment/nature-based-climate-solutions.html>

Government of Canada. (2023 Jan). Climate change funding programs. <https://www.canada.ca/en/services/environment/weather/climatechange/funding-programs.html>

Government of Canada. (n.d.). Powering Canada Forward: Building a Clean, Affordable, and Reliable Electricity System for Every Region of Canada. Foreword by Steven Guilbeault, Minister of Environment and Climate Change, and the Honourable Jonathan Wilkinson, Minister of Natural Resources. Author <https://natural-resources.canada.ca/our-natural-resources/energy-sources-distribution/electricity-infrastructure/powering-canada-forward-building-clean-affordable-and-reliable-electricity-system-for/25259>

Graney, E. (2022, Sep 23). Oilsands putting profits ahead of lower emissions, report says. *The Globe and Mail* (Ontario Edition). <https://www.pressreader.com/canada/the-globe-and-mail-ontario-edition/20220923/281913071977649>

Graney, E. (2023a, Jun 27). Canada's oil industry not prepared for drop in global demand: Study. *The Globe and Mail.* <https://www.theglobeandmail.com/business/article-canada-policy-oil-gas-decline/?login=true>

Graney, E. (2023b, Aug 16). CEO says Suncor too focused on energy transition, vows return to low-cost oil production, profitability. *The Globe and Mail* (Prairie Edition). <https://www.theglobeandmail.com/business/article-suncor-should-return-to-its-oil-production-roots-focus-on-fundamentals/>

Graney, E. (2023c, Oct 17). Suncor CEO explains sustainability comments, says company still committed to reducing emissions. *The Globe and Mail.* <https://www.theglobeandmail.com/business/article-suncor-ceo-greenhouse-gas-emissions-sustainability/>

Green Budget Coalition. (n.d.). Recommendations for budget 2023. Author. <https://greenbudget.ca/>

Gunton, T., Wheeler, D., & Tienhaara, K. (2023, Aug 2). Ottawa's plan to end fossil fuel subsidies doesn't actually end a single one. *The Globe and Mail.* <https://www.theglobeandmail.com/business/commentary/article-trudeau-fossil-fuel-subsidies/>

Guterres, A. (2022, April 4). UN climate report: It's "now or never" to limit global warming to 1.5 degrees. *UN News.* <https://news.un.org/en/story/2022/04/1115452>

Hall, E. T. (1976). *Beyond culture.* Anchor Books.

Handcock, A. (2022, Aug 18). Europe fires destroy record 700,000 hectares of forest. *Financial Times.* <https://www.ft.com/content/d3941a20-38af-483b-a1f7-9df844dcfb68#post-e103b129-cb3f-4a11-be8b-235df305cfcd>

Hansen, J. (n.d.). James Hansen. *Wikipedia.* <https://en.wikipedia.org/wiki/James_Hansen>

Harder, G. (2023, Aug 9). Years of drought wearing on farms. *Saskatoon StarPhoenix*. <https://thestarphoenix.com/news/local-news/saskatchewan-crops-pastures-suffering-from-drought-as-harvest-begins>

Hatcher, A., Bartlett, C., Marshall, A., & Marshall, M. (2009). Two-Eyed seeing in the classroom environment: Concepts, approaches, and challenges. *Canadian Journal of Science, Mathematics and Technology Education*, 9 (3), 141–153.

Hayes, A. (2022, June 22). What is economics? *Investopedia*. <https://www.investopedia.com/terms/e/economics.asp>

Heginbottom, J. A., Dubreuil, M. A., & Harker, P. T. (1995). GEOSCAN search results: Fastlink. *The national atlas of Canada* (MCR Series no. 4177). <https://doi.org/10.4095/294672>

Hendriks, F., Kienhues, D., & Bromme R. (2020). Replication crisis = trust crisis? The effect of successful vs failed replications on laypeople's trust in researchers and research. *Public Understanding of Science*, 29(3), 270–288. <https://doi.org/10.1177/0963662520902383>

Hendry, Joy. (2014). *Science and sustainability: Learning from Indigenous wisdom*. Palgrave, Macmillan.

History Staff. (2020, Aug 18). Did Nero really fiddle while Rome burned? *History*. <https://www.history.com/news/did-nero-really-fiddle-while-rome-burned>

Hotez, P. J. (2023, Sept 19). The deadly rise of anti-science: A scientist's warning. John Hopkins University Press. <https://www.press.jhu.edu/books/title/33293/deadly-rise-anti-science>

IPCC. (Intergovernmental Panel on climate change). (2023, Mar 20). Urgent climate action can secure a liveable future for all. (Press Release).

Israelson, D. (2023, Oct 30). Building Canada's cleaner and greener energy future. *The Globe and Mail*. <https://www.theglobeandmail.com/business/article-building-canadas-cleaner-and-greener-energy-future/?login=true>

James, W., Vaaldmanis, R., & Evans, D. (2022, Nov 9). Poor nations criticize Big Oil, wealthy countries at COP_{27} climate summit. *The Globe and Mail*. <https://www.theglobeandmail.com/business/international-business/article-cop27-summit-developing-nations-criticize-big-oil-wealthy-governments/>

Jones, J., & Bradshaw, J. (2022, Sep 29). Canadian banks having second thoughts about Mark Carney's green alliance over legal, governance risks. *The Globe and Mail*. <https://www.theglobeandmail.com/business/article-mark-carney-gfanz-banks/>

Jones, J., & Graney, E. (2023, Aug 5). Clean-energy companies rattled by Alberta moratorium. *The Global and Mail* (Prairie Edition). <https://www.theglobeandmail.com/business/article-alberta-risks-losing-billions-in-renewable-energy-investments-with/>

Kawagley, A. O. (1995). A *Yupiaq worldview*. Waveland Press.

Kenton, W. (2022, May 13) What is an economist? *Investopedia*. <https://www.investopedia.com/terms/e/economist.asp>

King, T. (2003). *The truth about stories*. House of Anansi Press.

Kleespies, M. W., & Dierkes, P. W. (2022). The importance of the sustainable development goals to students of environmental and sustainability studies—A global survey in 41 countries. *Humanities and Social Sciences Communications*, 9(article number 218), 1–9. <https://doi.org/10.1057/s41599-022-01242-0>

Klopfer, L. E., & Aikenhead, G. S. (2022). Humanistic science education: The history of science and other relevant contexts. *Science Education, 106*(3), 490–504.

Klopfer, L. E., & Cooley, W. W. (1961). *Test on understanding science.* Educational Testing Services.

Klopfer, L. E., & Cooley, W. W. (1963). "The History of Science Cases" for high school in the development of student understanding of science and scientists. *Journal of Research in Science Teaching, 1*(1), 33–47.

Kolstø, S. D. (2001). 'To trust or not to trust, . . .' —Pupils' ways of judging information encountered in a socio-scientific issue. *International Journal of Science Education, 23*(9), 877–901.

Kurz, L. (2023, Nov 22). Feds say 2035 target for clean energy possible in Sask. *Saskatoon StarPhoenix.* <https://thestarphoenix.com/news/feds-say-2035-target-for-clean-energy-possible-in-sask/wcm/602c2b9e-5d3d-4217-a53a-c5945477cd98>

Krishnamurthy, R. (2022, Nov 20). A new dawn for humanity? COP_{27} ends; loss and damage fund, other agendas adopted. *Down to Earth.* <https://www.downtoearth.org.in/news/climate-change/a-new-dawn-for-humanity-cop27-ends-loss-and-damage-fund-other-agendas-adopted-86087>

Layton, D. (1986). Science education and values education--An essential tension. In J. Brown, A. Cooper, T. Horton, F. Toates, & Zeldin (Eds.), *Science in schools* (pp. 110-120). Open University Press.

Lederman, N. G. (2007). Nature of science: Past, present, and future. In S. K. Abell & N. G. Lederman (Eds), Handbook of research on science education (pp. 831-880). Lawrence Erlbaum Associates.

Lee, H., Yen, C.-F., & Aikenhead, G. S. (2012). Indigenous elementary students' science instruction in Taiwan: Indigenous knowledge and Western science. *Research in Science Education*, 42(6), 1183–1199. <https://doi.org/10.1007/s11 165-011-9240-7>

Levy, B. (2023, Jul 24). Forecast offers little relief as drought withers Saskatchewan. *Saskatoon StarPhoenix.* <https://thestarphoenix.com/news/local-news/forecast-offers-little-relief-as-drought-withers-saskatchewanLinnebo>, Ø. (2018). Platonism in the philosophy of mathematics. In W. N. Zalta (Ed.), *The Stanford encyclopedia of philosophy.* <https://plato.stanford.edu/archives/spr2018/entries/platonism-mathematics/>

Lewis, A. (n.d.-1). They knew. They profited. They need to pay their fair share (website). <https://suebigoil.ca/>

Lewis, A. (n.d.-2). About Avi (website). Author. <https://avilewis.ca/about/>

Lewis, S. (2023, Jul 18). How we got to this point (podcast). CBC Ideas. Suzuki's survival guide. <https://www.cbc.ca/radio/ideas/david-suzuki-s-survival-guide-a-retrospective-1.6894102>

Linnebo, Ø. (2018). Platonism in the philosophy of mathematics. In W. N. Zalta (Ed.), The Stanford encyclopedia of philosophy. https://plato.stanford.edu/archives/spr2018/entries/platonism-mathematics/

Little Bear, L. pc. (personal communication) follows the oral tradition of Indigenous cultures. There need not be a date, title, or location. This applies for Elders, and experienced Knowledge Holders.

Little Bear, L. (2000). Jagged worldviews colliding. In M. Battiste (Ed.), *Reclaiming indigenous voice and vision* (pp. 77–85). UBC Press.

Little Bear, L. (2005). <https://www.youtube.com/watch?v=gJSJ28eEUjI>

Little Bear, L. (2011). Native and western science: Possibilities in a dynamic collaboration (YouTube). <https://www.youtube.com/watch?v=ycQtQZ9y3lc>

Lomborg, B. (2022, Oct 31). The climate hypocrisy of rich countries. *Globe & Mail.* <https://www.theglobeandmail.com/opinion/article-climate-summit-hypocrisy/>

Lux, A. (2021, Dec 21). Are you a more holistic or analytic thinker? *The Conversation.* <https://theconversation.com/are-you-a-more-holistic-or-analytic-thinker-take-this-quiz-to-find-out-170018>

Mackell, A. G. (2022, Nov 7). Climate conference COP$_{27}$: The sham in Sharm El Sheikh. Global Order. <https://www.democracywithoutborders.org/24944/climate-conference-cop27-the-sham-in-sharm-el-sheikh/>

MacLeod, R. (1981). Introduction. The advancement of science. In R. MacLeod & P. Eollins (Eds.), *The parliament of science* (pp. 17–42). Science Reviews.

Mandryk, M. (2023, May 19). Deadline a non-starter for SaskPower. *Saskatoon Starphoenix.* A URL is not available.

Marshall, A. (2007, May). *Two-Eyed seeing.* A presentation to the Two-Eyed Seeing Workshop for Science Education for Children and Youth. Cape Breton University, Nova Scotia, Canada. <http://marcatodigital.com/iish/pdf/may_24_workshop_summary.pdf>

Maryboy, N. C. (2011). Quoted in Aikenhead, G., & Michell, H. (2011, p. 198). *Bridging cultures: Indigenous and scientific ways of knowing nature.* Pearson Canada.

Maryboy, N. C., & Begay, D. H. (2005). *Sharing the skies: Navajo astronomy—A cross-cultural view.* Bluff, UT: Indigenous Education Institute.

Maryboy, N. C., Begay, D. H., & Nichol, L. (2006). Paradox and transformation. In World Indigenous Nations Higher Education Consortium (Ed.,), *INDIGENOUS voices, indigenous places?* Te Tauihu o Nga Wananga, New Zealand. <https://www.worldcat.org/title/456555027>

Mason, G. (2023, Aug 18). The clean energy shift is happening whether the UCP [United Conservative Party] likes it or not. *The Globe and Mail* (Prairie Edition). <https://www.theglobeandmail.com/opinion/article-the-clean-energy-shift-is-happening-whether-albertas-government-likes/>

McGreal, C. (2021, Jun 30). Big oil and gas kept a dirty secret for decades. Now they may pay the price. *The Guardian.* <https://www.theguardian.com/environment/2021/jun/30/climate-crimes-oil-and-gas-environment>

McKenna, C. (2023, Sept 20). Oil giants' climate-change cheap talk is why we need a hard cap on their emissions. *The Globe and Mail* (Report of Business). <https://www.theglobeandmail.com/business/commentary/article-oil-companies-emissions-climate-change/>

Medin, D. L., & Bang, M. (2014). *Who's asking? Native science, western science, and science education.* The MIT Press.

Meyer, S. (n.d.). Kôhkum's Gathering. <http://sites.google.com/nesd.ca/kohkumsgathering/home>

Meyer, S., Aikenhead, G., Cardinal, K., et al. (2019). *Indigenous culture-based school mathematics for reconciliation and professional development* (research report). Saskatoon, Saskatchewan, Canada. Stirling McDowell Foundation. <https://mcdowellfoundation.ca/culture-based-school-mathematics-for-reconciliation-and-professional-development/>

Michell, H. J. (2011). *Working with elders and indigenous knowledge systems: A reader and guide for places of higher learning.* JCharlton Publishing Ltd.

Michell, H. J. (2013). *Cree ways of knowing and school science.* JCharlton Publishing Ltd.

Michell, H. J. (2017). *Reconciliation from an indigenous perspective: Weaving the web of life in the aftermath of residential schools.* JCharlton Publishing Ltd.

Michell, H., Vizina, Y., Augustus, C., et al. (2008). *Learning indigenous science from place.* Aboriginal Centre, University of Saskatchewan.

Middleton, W. E. K. (1963). *The scientific revolution.* Canadian Broadcasting Corporation Publications.

Mistry, J., & Berardi, A. (2016). Bridging indigenous and scientific knowledge. *Science, 352*(6291), 1274–1275. <http://dx.doi.org/10.1126/science.aaf1160>, <https://oro.open.ac.uk/cgi/search/archive/unified?screen=Search&dataset=archive&q_merge=ANY&q=Bridging+indigenous+and+-scieentificknowledge&_action_search=Search&order=-date%2Fcreators_name%2Ftitle#>

NASA. (2017). Johannes Kepler: His life, his laws and times. Author. <https://www.nasa.gov/kepler/educat ion/joh

N.A.S.A. (2017). Johannes Kepler: His life, his laws and times. Author. <https://www.nasa.gov/kepler/educat ion/johan nes>

Neis, K. (2023, Aug 16). Sask. and federal NDP at odds over net-zero. *Saskatoon StarPhoenix*, p. A1. URL not available.

Nickel, R. (2023, Aug 21). Renewable power firms hit brakes on Alberta projects after UCP pauses approvals. *CBC News.* <https://www.cbc.ca/news/canada/calgary/renewables-pause-alberta-projects-companies-1.6942597>

Nilsen, E. (2030, Aug 14). Montana judge hands young plaintiffs significant victory in landmark climate trial. *CNN.* <https://www.cnn.com/2023/08/14/us/montana-kids-win-climate-trial>

NPR staff. (2019, Sept 23). Transcript: Greta Thunberg's speech at the U.N. Climate Action Summit. *npr* (National Public Radio). <https://www.npr.org/2019/09/23/763452863/transcript-greta-thunbergs-speech-at-the-u-n-climate-action-summit>

O'Connor, F. M., Boucher, O., Gedney, N., et al. (2010). Possible role of wetlands, permafrost, and methane hydrates in the methane cycle under future climate change: A review. *Reviews of Geophysics*, 48(RG4005), 1–33. <https://doi.org/10.1029/2010RG000326>

OECD. (2023). *Climate tipping points: A critical moment for action.* Author. Climate Tipping Points: Insights for Effective Policy Action.

Orange, A. D. (1981). The beginnings of the British Association. 1831–1851. In R. MacLeod & P. Collins (Eds.), *The parliament of science.* Science Reviews.

Osborne, J., & Pimental, D. (2022). Science, misinformation, and the role of education. *Science*, *378*(6617), 246–248. <https://doi.org/10.1126/science.abq8093>

Pearson Canada (textbook series of 7 books for grades 3–9). (2009–2012). *Pearson Saskatchewan Science.* Authors: (multiple for each textbook). Reid McAlpine.

Popp, J. (2018, Jan 2). How Indigenous knowledge advances modern science and technology. *The Conversation.* <https://theconversation.com/how-indigenous-knowledge-advances-modern-science-and-technology-89351>

Rabson, M. (2023, Aug 5). Extreme weather risks changing Canada's insurance industry. *The Global and Mail* (Prairie Edition). <https://www.theglobeandmail.com/canada/article-extreme-weather-risk-changing-canadas-insurance-industry-raising-costs-2/>

Randall Anthony Communications. (2022, Nov 18). Excellence in research & innovation. *The Globe and Mail*. pp. CFI 1–CFI 21.

Reed, M. (2022, Nov 18). Conservation, learning, reconciliation (Sponsor content, pp. SFI 1 to SFI 21). *The Globe and Mail*. <https://www.theglobeandmail.com/life/adv/article-conservation-learning-reconciliation/?login=true>

Reguly, E. (2022a, Nov 14). At COP_{27}, few embrace the concept of loss and damage. *The Globe and Mail*. <https://www.theglobeandmail.com/business/international-business/article-at-cop27-few-embrace-the-concept-of-loss-and-damage/>

Reguly, E. (2022b, Nov 19). Climate talks at COP_{27} extended through Saturday in attempt to break deadlock. *The Globe and Mail*. <https://www.theglobeandmail.com/business/commentary/article-cop27-climate-talks-deadlock/>

Reis, G. (2022). Education for sustainability and environment: Searching for the perfect plate. *Science & Education, 31*(3), 851–854. <https://doi.org/10.1007/s11191-021-00273-4>

Routley, N. (2022, Jul 20). 5 things to know about Europe's scorching heatwave. Visual Capitalist: Environment. <https://www.visualcapitalist.com/5-things-to-know-about-europes-scorching-heatwave/>

Royal Museums Greenwich. (n.d.). Follow the NASA Artemis missions. Author. <https://www.rmg.co.uk/stories/topics/nasa-moon-mission-artemis-program-launch-date>

Rudolph, J. L. (2005). Epistemology for the masses: The origins of "The Scientific Method" in American schools. *History of Education Quarterly, 45*(3), 341–376.

Saba. R. (2023, Aug 26). As wildfire risks grow, insurers feeling the strain. *Saskatoon StarPhoenix*.

Sadeghi, M. (2023, Aug 14). Fact check: A 1912 article about burning coal and climate change is authentic. USA Today. <https://www.usatoday.com/story/news/factcheck/2021/08/13/fact-check-yes-1912-article-linked-burning-coal-climate-change/8124455002/>

Sanders, R. (2022, October 21). Canada is 'weaving' Indigenous science into environmental policy-making. CBC News. <https://www.cbc.ca/news/science/what-on-earth-indigenous-science-environment-canada-1.6623377>

Saunders, D. (2023, Jul 29). This time, climate change is personal. *The Global and Mail* (Prairie Edition). Opinion section, p. 011. Science Museum (London). (2019, Aug 9). The 17th century society that transformed science. Author. <https://www.sciencemuseum.org.uk/objects-and-stories/17th-century-society-transformed-science>

Semeniuk, I., & Stueck, W. (2023, Mar 20). Actions to stave off climate breakdown are possible but urgent, UN report says. *The Globe and Mail*. <https://www.theglobeandmail.com/business/article-actions-to-stave-off-climate-breakdown-are-possible-but-urgent-un/>

Shingler, B. (2023, Jun). Who is liable for the climate crisis? Canadians are going to court to find out. CBC *News*. <https://www.cbc.ca/news/climate/canada-climate-litigation-montana-1.6879410>

Simakov, A. (2023, Aug 5). Pause on solar, wind projects risk jobs. *The Global and Mail* (Prairie Edition). <https://www.theglobeandmail.com/business/commentary/article-albertas-pause-on-solar-and-wind-projects-risks-jobs-investment-and/>

Solorio, I., Romero, R., & Ros, C. (2023). Indigenous and local knowledge in environmental decision making. In H. Jörgens, H. C. Knill, & Y. Steinebach (Eds.), *Routledge handbook of environmental policy* (pp. 194–206). Routledge. ISBN 9780367489922.

Sriraman, B. (Ed.). (2017). *Humanizing mathematics and its philosophy*. Springer International Publishing.

Stephenson, A. (2022, Nov 10). Pathways president says industry will be judged on whether it accomplishes its goals. *The Canadian Press*. <https://globalnews.ca/news/9268034/pathways-alliance-oilsands-cop27-climate-change/>

Stokel-Walker, C. (2022, Oct 19). Hurricane Ian destroyed their homes. Algorithms sent them money. *Business*. <https://www.wired.com/story/hurricane-ian-destroyed-homes-google-algorithms-sent-money/>

Stout, L. A. (2002). The investor confidence game. *Brooklyn Law Little Bear Review*, 68(2), 407–437. <http://scholarship.law.cornell.edu/facpub/762>

Stueck, W. (2024, Jan 10). With 2023 hottest year on record, the world is now on track to eclipse 1.5 C above preindustrial levels. *The Globe and Mail*. <https://www.theglobeandmail.com/canada/article-with-2023-hottest-year-on-record-the-world-is-now-on-track-to-eclipse/?login=true>

Suldovsky, B., Landrum, A., & Stroud, N. J. (2019). Public perceptions of who counts as a scientist for controversial science. *Public Understanding of Science*, 28(7), 797–811. <https://doi.org/10.1177/0963662519856768>

Suzuki, D. (2015). *Letters to my grandchildren*. Greystone Books.

Suzuki, D. (2023a, Jul 11). The warning (podcast). CBC Ideas. Suzuki's survival guide. <https://www.cbc.ca/radio/ideas/david-suzuki-s-survival-guide-a-retrospective-1.6894102>

Suzuki, D. (2023b, Jul 18). How we got to this point (podcast). CBC Ideas. Suzuki's survival guide. <https://www.cbc.ca/radio/ideas/david-suzuki-s-survival-guide-a-retrospective-1.6894102>

Tait, C., & Graney, E. (2023, Aug 7). Politics behind Alberta decision on wind, solar power expansion, say critics. *The Globe and Mail* (Prairie Edition). <https://www.theglobeandmail.com/canada/alberta/article-politics-behind-alberta-decision-on-wind-solar-power-expansion-say/>

Talaga, T. (2018). *All our relations: Finding the path forward*. House of Anansi Press.

Tank, P. (2022, Oct 13). Sask. climate costs of $111 billion a wildly misleading sum. *Saskatoon StarPhoenix*. <https://thestarphoenix.com/opinion/columnists/phil-tank-sask-climate-costs-of-111-billion-a-wildly-misleading-sum>

The Canadian Press. (2021, May 22). Film producer Avi Lewis named as NDP candidate in B.C. riding. The *National Post*. <https://nationalpost.com/news/canada/film-producer-avi-lewis-to-seek-ndp-nod-in-b-c>

Tiner, J. H. (1977). *Johannes Kepler: Giant of faith and science*. Mott Media.

Toronto Conference. (n.d.). Wikipedia. <https://en.wikipedia.org/wiki/Toronto_Conference_on_the_Changing_Atmosphere#cite_note-Gordon_Suzuki_1990-14>

Truth and Reconciliation Commission. (2016). *A knock on the door.* University of Manitoba Press.

Tsuji, L J. S., & Ho, E. (2002). Traditional environmental knowledge and western science: In search of common ground. *Canadian Journal of Native Studies*, *2*(2002), 327–360.

Vanderclippe, N. (2013, Sep 21). Climate change may mean "hundreds of billions" at risk in U.S. housing market. *The Globe and Mail.* <https://www.theglobeandmail.com/world/article-climate-change-may-mean-hundreds-of-billions-at-risk-in-us-housing/?login=true>

Venkatesh, B. (2022, Nov 18). Driving societal transformation (Sponsor Content, p. CFI 4). *The Globe and Mail.*

Wallace-Wells, D. (2020). *The uninhabitable earth: Life after warming.* Tim Duggan Books.

Weiss, K., Hamann, M., & Marsh, H. (2013). Bridging knowledges: Understanding and applying indigenous and western scientific knowledge for marine wildlife management. *Society & Natural Resources*, *26*(3), 285–302. <https://doi.org/10.1080/08941920.2012.690065>

Wikipedia. (n.d.). Toronto conference on the changing atmosphere (website). <https://en.wikipedia.org/wiki/Toronto_Conference_on_the_Changing_Atmosphere>

Wilkinson, J. (2023, May 19). Net-zero power for Sask. An economic imperative. *The Globe and Mail* (Prairie Edition). <https://thestarphoenix.com/opinion/columnists/jonathan-wilkinson-sask-premier-is-wrong-about-net-zero-grid-by-2035>

Willis, A. (2022, Dec 5). TC energy talks resiliency in the transition to low carbon. *The Globe and Mail.* <https://www.theglobeandmail.com/business/article-tc-energy-ceo-talks-resiliency-in-the-transition-to-low-carbon/?login=true>

Wood, S. (2023, Sep 24). Mathur V Ontario: Grounds for optimism about recognition of a constitutional right to a stable climate system in Canada. *McGill Law Journal*, *69*, 1–43. <https://ssrn.com/abstract=4591439>

World Meteorological Organization. (2023, Jun 19). "Climate change impacts scar Europe", but increase in renewables signals hope for future Press Release Number: 19062023. Author. <https://public.wmo.int/en/media/press-release/climate-change-impacts-scar-europe-increase-renewables-signals-hope-future>

Yeo, R. (1981). Scientific method and the image of science, 1831–1890. In R. MacLeod & P. Collins (Eds.), *The parliament of science* (pp. 65–88). Science Reviews.

Younging, G. (2018). *Elements of indigenous style: A guide for writing by and about indigenous peoples*. Brush.

Zeidler, D. L., Walker, I. A., Ackett, W. A., et al. (2002). Tangled up in views: Beliefs in the nature of science and responses to socioscientific dilemmas. *Science Education*, *86*(3), 343–367.

Zeyer, A. (2023). They always pretend to have the whole truth: How the epistemization of politics endangers democracy, and what this may have to do with scientism in school science. *Science Education*, *107*(2), 275–290.

Ziman, J. (1998). Why must scientists become more ethically sensitive than they used to be? *Science*, *282*, 1813.

INDEX

ABOUT THE AUTHOR

(Glen Aikenhead)

Glen Aikenhead has a B.Sc. (Hons) in chemistry and physics from the University of Calgary (1965), and an MA from Harvard (1966). He enjoyed three years of teaching high school science and math before embarking on a Doctorate of Science Education at Harvard (1972), during which he established his life-long research program: humanistic school science for everyday life, defined by students' real-life experiences. He took up a full professorship at Saskatchewan's College of Education in 1980. About a decade later, he made a mid-course correction to his research program by giving increased emphasis to the unique cultural challenges faced by Indigenous students in Western science and math classes. Guided by Elders, he learned to interpret the world from the perspective of Indigenous wisdom. Professor Aikenhead is the author of 10 academic books, and has served as visiting professor at institutions around the world, including Tokyo University of Science, Japan; Kristianstad University, Sweden; the University of Lisbon, Portugal; Providence University, Taiwan; and the University of Waikato, Aotearoa New Zealand.

www.ingramcontent.com/pod-product-compliance
Ingram Content Group UK Ltd.
Pitfield, Milton Keynes, MK11 3LW, UK
UKHW020842290926
15470UKWH00004B/1498

9 781636 678054